ES&T Presents

VCR
Troubleshooting and Repair

ES&T Presents
VCR
Troubleshooting and Repair

PUBLICATIONS

©1999 by Electronic Service Technology

PROMPT© Publications is an imprint of Sams Technical Publishing, 5436 W. 78th Street, Indianapolis, IN 46268.

All rights reserved. No part of this book shall be reproduced, stored in a retrieval system, or transmitted by any means, electronic, mechanical, photocopying, recording, or otherwise, without written permission from the publisher. No patent liability is assumed with respect to the use of the information contained herein. While every precaution has been taken in the preparation of this book, the author, the publisher or seller assumes no responsibility for errors or omissions. Neither is any liability assumed for damages resulting from the use of information contained herein.

International Standard Book Number: 0-7906-1158-9
Library of Congress Catalog Card Number: 98-065745

Acquisitions Editor: Loretta Yates
Editor: J.B. Hall, Nils Conrad Perrson
Assistant Editors: Pat Brady
Contributing Editors: Steve Babbert, Homer L. Davidson, Timothy W. Durhan, Wayne B. Durham, Juergen Ewert, Arthur Flavell, T. V. Kappel, Victor Meeldijk, Lamar Ritchie ES&T Staff, WG, Philip Zorian
Typesetting: J.B. Hall
Indexing: J.B. Hall
Cover Design: Christy Pierce
Graphics Conversion: Terry Varvel
Illustrations and Text: Supplied by *Electronic Servicing & Technology* (*ES&T*) Magazine, CQ Communications, Inc., 76 N. Broadway, Hicksville, NY 11801.

Trademark Acknowledgments:
All product illustrations, product names and logos are trademarks of their respective manufacturers. All terms in this book that are known or suspected to be trademarks or services have been appropriately capitalized. PROMPT® Publications and Sams Technical Publishing cannot attest to the accuracy of this information. Use of an illustration, term, or logo in this book should not be regarded as affecting the validity of any trademark or service mark.

PRINTED IN THE UNITED STATES OF AMERICA

9 8 7 6 5 4 3 2

Table of Contents

Chapter 1
Inspection, Cleaning and Lubrication of Camcorders
By the ES&T Staff   1
Required Maintenance   1
Scheduled Maintenance   2
Check before starting repairs   2
Tools and Materials Needed For Camcorder Inspection and Maintenance   3
Cleaning the Video Heads   3
Cleaning the Tape Transport Components   4
Oiling   4

Chapter 2
VCR Troubleshooting Tips
By Victor Meeldijk   5
Frequent VCR Failures   5
Troubleshooting By Symptom Analysis   6
Picture Problems (General)   6
Picture Out of Sync   6
Snow Bar or Alternating Clear/Snowy Picture   7
Noisy Picture or Absence of Video   8
Flagging   9
No Picture In TV Mode   10
No Picture In Stop Mode   10
Color-related Picture Problems   10
Audio Problems   11
Audio and Video Problems   13
Tape Interchange Problems   14
VCR Control Problems   14
Miscellaneous Symptoms   17

Chapter 3
Video Update: Setting VCR Head Switching
By the ES&T Staff   19
Why Head Switching Needs Adjustment   19
Adjusting Head Switching By Counting Pulses   21
Using a Dual-trace Scope With Delta Time Capability   22
Using Delta Time to Adjust Head Switching   24

Table of Contents

Chapter 4
Movable Heads Drive Circuitry
By the ES&T Staff   *27*
Block Diagram Analysis   27
Head Selection and Signal Detection Circuitry   28
MV Heads Drive Signal Generation   30
MV Drive Amplifier Circuitry   32
Protect Circuitry   34
Troubleshooting Tips   34

Chapter 5
Camcorder Electrical Adjustment
By the ES&T Staff   *37*
Digital Adjustments   37
The EEPROM   38
Two Ways To Make Adjustments   39
Other Instruments Required   39
Hexadecimal Representation   39
More To Come   40

Chapter 6
The Gentle Art of Camcorder Repair
By T.V. Kappel   *41*
Evaluating the Damage   41
The Case Tells a Tale   42
Damage Behind the Lens   42
Back to the Operating Room   43

Chapter 7
VCR Servicing: Taking the Mystery Out of Video Head Replacement
By Philip M. Zorian   *45*
The Number of Video Heads   46
Diagnosing the Problem   46
Visual Inspection   47
Cleaning the Video Heads   47
Check the Pause Function and the Outputs   47
Check for Connection Problems   48
Tape Speed   48
Tape Tension   49

Table of Contents

Guidepost Height	49
Speed of Rotation of the Video Drum	50
Replacing the Video Heads	50
Installing the Replacement Drum	51
Conclusion	51

Chapter 8
Mechanical Problems in the Sanyo VHR9300
By Steve Babbert   53

The Broken Front-load Gear	53
Stuck Tape	54
Fast Forward/Rewind Problems	56
Summary	59

Chapter 9
Readout and Tape Loading Problems in RCA/Hitachi/Sears VCRs
By Victor Meeldijk   61

Models With Discrete Components	62
A Tape-stage Mechanism Problem	65
Solving the Tape-stage Problem	67
Aligning the Gears	68
Cosmetic Fixes	69

Chapter 10
Recommended System and Servo Control Circuit Diagnostic Procedures For VCRs
By the ES&T Staff   71

A Recommended Diagnostic Procedure	72
Some Diagnostic Hints	72
System Control Circuit Diagnosis	73
Servo Control Circuits	74
Diagnosing a Servo Control Circuit Problem	75

Table of Contents

Chapter 11
Servicing the Hitachi VM Series Camcorder
By Timothy W. Durhan   *77*

Symptoms of Worn Rubber Parts	77
Getting Started	78
Performing the Diagnosis	78
Getting to the Belts	79
Replacing the Belts	80
The Finishing Touches	81
Perform a Thorough Operational Test	82

Chapter 12
Servicing VCR Motor Problems
By Homer L. Davidson   *83*

Motor Supply Sources	84
Loading Motor Problems	85
Loading Motor Related Symptoms	86
RCA VLP900—No Loading Action	87
Capstan Motor Problems	88
Capstan Motor Related Problems	88
RCA VLT650HF—No Cylinder or Capstan Movement	89
Cylinder or Drum Motor Problems	89
Cylinder Motor Related Problems	90
RCA VLP900— No Cylinder Motor Rotation	90
Conclusion	91

Chapter 13
Solving a VCR Short Circuit
By Steve Babbert   *93*

Erratic Operation	93
Someone Else Had Been Here Before	94
The Problem Reappears	95
Correcting the Problem	97

Chapter 14
Solving VCR Servo System Problems
By Arthur Flavell   *99*

VHS Format	99
Servo Operation During Playback	100
Servo Operation During Record	102
Troubleshooting	103

Troubleshoot in This Sequence	104
Breaking the Loop	104

Chapter 15
VCR Mechanical Problems
By Philip Zorian — *107*

Replacing Belts; Don't Measure Them	107
The Ten Most Common Problems	108
Incomplete Loading	109
Eating the Tape	109
Tape Slippage	110
No Rewind	110
Squealing Sounds	110
Take-up Reel Won't Turn	111
Intermittent Shutoff	111
Tape-edge Damage	112
Dead VCR	112
Conclusion	113

Chapter 16
VCR Servo Problems: The Diagnostic Device Revisited—Part 1
By Steve Babbert — *115*

The Helical Scan System	116
The Servo System	116
The Speed Control Loop	117
The Phase Control Loop	117
The Tracking System	118
Head Drum Speed and Phase Control	118
The Auto Speed Select Circuit	118
Solving a Capstan Speed Problem	119
An Attempt With No Schematic	119
Checking For a Missing Signal	121
Opening the Loop	121
Examining the Synchronizing Pulses	122
Obtaining a Schematic	123
Solving the Problem	124
The Problem Solved	125

Table of Contents

Chapter 17
VCR Servo Problems: The Diagnostic Device Revisited—Part 2
By Steve Babbert   **127**

The Servo Circuit	127
The PG/FG Separator	128
Error Voltages	129
Solving a Servo Problem	129
The Strobe Method	130
Scoping the PWM Outputs	130
Opening the Loop	131
Using the Diagnostic Device	131
Adjusting the Cylinder Motor Speed	132
What the Tests Showed	133
The Low-pass Filter	133
Analysis of the Problem	134

Chapter 18
Where Do I Begin?: Analyzing VCR and Camcorder Problems
By Steven Jay Babbert   **137**

The Test Jig	137
Reel Motion Sensor	138
Watch the Take-up Reel	139
Idler Replacement	140
Pinch Roller Problems	140
System Control Problems	141
When the Unit Blows Fuses	142
Motor Control Circuits	142
Mechanical Alignment	143
The rf Modulator	144
Detached F-con	144
The Vacuum Fluorescent Display	145
Remote Control Problems	146
Syscon	146
Cleaning and Lubrication	146
Summary	147

Table of Contents

Chapter 19
VCR Service Centers: Tips on Remaining Profitable
By Wayne B. Graham   *149*
The Arithmetic of VCR Servicing	149
Charge For Estimates	149
Provide a Knowledgeable Accurate Estimate	150
Provide a Six-month To One-year Warranty on VCRs You Service	151
Make Money On Abandoned VCRs	151
Find More Business	152
Expensive Products Will Be Serviced	153

Chapter 20
Camcorder Servicing
By the ES&T Staff   *155*
The Sensor	155
The CCD	156
Troubleshooting the Camcorder	157
Camcorder Doesn't Operate At All	157
Tracking Down Power Problems	158
Divide and Conquer	159

Chapter 21
Hitachi VCR Repairs
By Victor Meeldijk   *161*
Realigning the VCR	162
Repairing a Hitachi Model VT-M250A	163

Chapter 22
Magnetic Recording Principles
By Lamar Ritchie   *165*
Video Recording	165
Frequency Translation	165
Recording the Audio	166
The Technology	166
Professional Video Recording	167
Helical Scan	167
U-Matic	170
Recording Chroma and Luma	171
Overcoming the Problems	172

Table of Contents

The Tension Regulator	172
The Dropout Compensator	172
Head-to-tape Contact	172
Getting the Signal To and From the Heads	173
Speed Control	173
Modulation and Demodulation	174
Mechanical Differences	174
Improved Formats	175
Compatibility Considerations	177
The Favorite	177
The 8-mm Format	177
Coming Attractions	178

Chapter 23
Magnetic Recording Principles: Video
By Lamar Ritchie   179

Azimuth Recording	179
Beat Frequencies	181
FM Interleaving	181
The Delay Device	183
The Playback Signals	184
The Head Preamp	185
The Servo Circuits	186
The Capstan Reference Generator	187
Motor Drive	188
Precise Tracking	188
The System Controller	188
Switches and Sensors	190
The Dew Sensor	190
The Reel Sensor	190
End of Tape Sensor	191
Tuner and Demodulator	192
RF Switching	192
Performance Options	193
Four Heads Also Eliminate Noise	193
The Three-head VCR	194
Many Variations	194

Table of Contents

Chapter 24
Magnetic Recording Principles: Audio and Video
By Lamar Ritchie   *195*
The Principles of Magnetic Recording   195
Audio Recording Requires a Bias Signal   195
The Magnetic Recording Head   197
The Head Nonlinearity Problem   198
Limitations in the Recording Process   199
Multiple Tracks   200
Improvements in Fabrication of the Tape   201
Video Tape Recording   201

Chapter 25
Dynamic VCR Head Check
By WG   *203*
Try This Dynamic Test   203
Setting Up the Equipment   204
Performing the Check   205
Most Inductors Should Work   205
Take Care in Constructing and Using the Coil   206
Comments   207
Use the Coil to Check All the Heads   207
Experimenting With Frequencies and Positioning   208

Index   211

Chapter 1
Inspection, Cleaning and Lubrication of Camcorders
By the ES&T Staff

This article is adapted from the service manual for the Hitachi VM-E35A/A(PX) 8mm video camera/recorder.

Camcorders go where no self-respecting piece of delicate consumer apparatus should go. People take them out when the temperature is well below zero. They put them in the trunk of the car when the temperature outside is above 100 degrees, and the temperature in the trunk would be hot enough to barbecue a slab of ribs. All those videos of people at the beach mean that the videographer has exposed the camcorder to sun, sea and sand. I even saw one camcorder waiting on the shelf of a service center that had been taken to a monster truck rally. The videographer got some great footage, but the outside of his camcorder had been caked with flying muck. The service manager suggested to the owner that next time he might want to put his camcorder in a transparent plastic bag, but for this time he had on his hands a camcorder that needed some serious cleaning. Given the abuse that camcorders are exposed to, it's no wonder that they need maintenance, cleaning and lubrication, and a thorough once over every so often.

Required Maintenance

Here is a recommended maintenance schedule and procedure for camcorders that are used routinely. Naturally, you should recommend to your customers that if they abuse their camcorders that they should have them checked more often. The cleaning/lubrication discussed here is based on an 8mm video camcorder, but is applicable to all camcorders. The tape transport system in an 8mm camcorder uses very precise components. If any of these components is worn or dirty, the symptoms will be the same as if the component is defective. To ensure a good picture, the tape transport system must be cleaned and lubricated periodically, and worn-out components must be replaced.

1 *Inspection, Cleaning and Lubrication of Camcorders*

Scheduled Maintenance

Schedules for maintenance and inspection are not fixed because they vary greatly according to the way in which the customer uses the camcorder, and the environment in which it is used. In general home use, a good picture will be maintained if inspections are done every 1,000 hours or so. *Table 1-1* shows an inspection schedule based on the number of hours the camera is used each day.

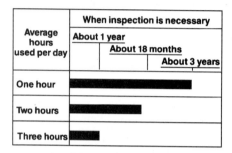

Table 1-1. Recommended maintenance intervals for camcorders depend on the amount of usage the product gets. Of course, if a camcorder is abused by being exposed to dust, moisture or excessive heat, maintenance intervals should be shorter.

Check Before Starting Repairs

The faults shown in *Table 1-2*, which appear during playback, may be remedied in many cases with just a cleaning and lubrication of the tape transport mechanism. When a camcorder is brought in to you with one or more of these problems, we suggest that you perform the indicated maintenance procedures and check the results before you begin any other service procedure.

Symptom	Inspection location
Poor picture S/N, No color	Dirt on video head or worn video head
Tape does not run or tape is slack	Dirt on pressure roller, in tape transport system or/and on cylinder
Vertical jitter	Dirt on video head or in tape transport system
Low volume or distorted sound	Dirt on video head or worn video head

Table 1-2. A dirty tape transport system can cause problems in a camcorder. If the owner of a camcorder complains of symptoms, try cleaning and lubing the unit before you perform any diagnostic procedures.

Tools and Materials Needed For Camcorder Inspection and Maintenance

If you plan to perform camcorder and/or VCR service on a regular basis, you should have all of the following materials on hand (see *Table 1-3* for the applications):

- Head cleaning kit
- VCR oil kit
- Ethyl alcohol (ethanol)
- Lint-free gauze
- Cleaning tape (available from many suppliers, including 3M and Maxell).

Name	Using locations
Sonic Slidas Oil (#1600)	Lubricate low-speed rotating sections
Froil (G31-SAY)	Grease metal components or molded sections subject to light load
Lock paint	Fix adjustment screws and nuts

Table 1-3. Use this information to determine what lubricant to use for the various portions of the camcorder tape transport.

Cleaning the Video Heads

The first step in cleaning the video heads is to run a cleaning tape through the transport. Consult manufacturer's literature, information from the manufacturer of the tape, and your distributor to determine the applicability of any particular tape. Before using it, carefully read the instructions that come with it. If the cleaning tape doesn't remove all of the oxide and other contaminants on the heads and head drum, use the head cleaning kit. Follow this procedure (see *Figure 1-1*):

- Moisten the chamois leather on the cleaning stick with cleaning fluid.
- Touch the moistened chamois leather on the cleaning stick against the head tip, and gently turn the upper cylinder back and forth against the leather.

1 *Inspection, Cleaning and Lubrication of Camcorders*

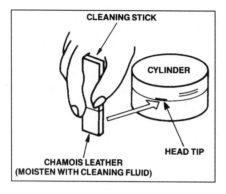

Figure 1-1. If the head cleaning tape doesn't completely clean the head and the head drum, use a chamois-tipped swab moistened with cleaning fluid, as shown here and as described in the text. Never move the swab vertically against the head, and never touch the head with any part of the swab except the chamois tip, as this could cause damage to the head.

Caution: Do not move the cleaning stick vertically against the head. This could damage the head. Also, make sure that only the chamois leather portion of the stick touches the head. If any other part of the stick touches the head, the head could be damaged.

Cleaning the Tape Transport Components

To clean the tape transport components, moisten some gauze or VCR cleaning foam swabs, or other manufacturer-recommended materials with ethyl alcohol, and use it to wipe the surfaces of transport system components that come into contact with the tape. Use extreme care to make sure that the tape transport components are not damaged or deformed by cleaning.

Oiling

Use only manufacturer-recommended lubricants in the camcorder. Use an oiler to apply one or two drops of Sonic Slidas oil to the specified components. Refer to the manufacturer's service manual for lubricating locations.

Caution: Do not use any more oil than recommended. Any excess oil may spill over or leak and come into contact with rotating parts. This could cause slippage or other defects. If you accidentally apply too much oil, wipe it clean with gauze slightly moistened with ethyl alcohol. Oiling should be done about every 1,000 hours of use.

Chapter 2
VCR Troubleshooting Tips
By Victor Meeldijk

In a study conducted by the Electronics Industries Association (EIA), 43% of five-year old VCR's have required service. Of the two most often cited reasons for service, head cleaning was the first (30% of units brought in for service) and rewind mechanism problems was the second (15%). The owner of a service center in New York estimated that 70% of his VCR service involved mechanical parts. The other 30% was attributed to electrical problems. This article discusses the reasons for VCR problems, and describes how to narrow down the problem areas by analyzing the failure symptoms.

Frequent VCR Failures

A high percentage of VCR repairs involves mechanical parts. One typical manufacturer's 10 most likely causes of warranty failure are:

- Belts 21.5%
- Idlers and pulleys 4.4%
- Motors 2.5%
- Switches 1.5%
- Loading mechanism 1.6%
- Relays 1.0%
- Tape heads 4.6%
- Circuit card assemblies 3.9%
- IC's 8.8%
- Remote unit 1.3%

Beyond the warranty period, mechanical wearout is one of the biggest problems along with slow rewind problems, belt failures, head clogging and physical VCR damage. In cases of electronics failures, the microprocessors are cited as the high failure item.

Troubleshooting By Symptom Analysis

Analyzing the symptoms of a malfunctioning VCR can often help in isolating the cause of the problem. The remainder of this article provides possible causes and suggested areas to check. They are listed by general, then specific, problem areas.

Picture Problems (General)

Remember that proper playback requires that both the position and speed of the video heads across the tape, and tension and position of the tape as it is fed past the heads, are correct. Following are some typical picture problems, possible causes, and areas to check.

Picture Out of Sync

An out of sync picture is characterized by rolling/jittering or noise bar running vertically (up and down) through the picture, non-hi-fi audio good. The audio quality depends on the rate at which the tape is pulled by the capstan across the audio head. If the non-hi-fi sound is good, then you know that tape speed is correct, so check the following areas in the servo circuit:

- Pulse width modulation signal
- Capstan frequency generator signal
- Cylinder pulse generator signal (a constant frequency signal based upon cylinder movement)
- Control track logic pulses
- 3OHz Reference Signal (29.97Hz square wave)

One way to check servo lock is to mark the top of the cylinder (sometimes referred to as the scanner) with a grease pencil. Use a fluorescent light, which will act as a strobe, to look at the cylinder while it is spinning. If the pencil mark does not appear as an almost stationary blurred pattern (in other words, if the mark is spinning around) it means that the servo is not locked.

If the servo locks in record, but not playback, this is a timing reference problem; check the tracking circuit. If lock doesn't occur in either record or playback, check the 30Hz feedback pulse from the video cylinder. Note: it may be difficult to see the pencil mark, so another way to perform this test is to note some reference mark on the top of the cylinder, such as a screw, or hole, and see if that reference appears to stand still under the fluorescent light.

Snow Bar or Alternating Clear/Snowy Picture

If the problem is either a snow bar, or an alternating clear/snowy picture, check the following areas.

- Tracking control settings
- Capstan phase: the video heads may be tracing the wrong azimuth track or the guard band
- Video head relay (or switching circuit). The heads may not be switching in or out of the circuit
- Back (tape) tension (especially if there are a lot of drop outs)

If in SP mode the top portion of the picture is clear, or in EP mode the bottom portion of the picture is clear, the heads are not the problem (provided the alignment potentiometers are not so misadjusted as to cause worn heads to track in such a way as to cause a portion of the picture to be clear as in *Figure 2-1*) and the video head relay has to be checked.

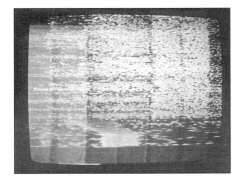

Figure 2-1. A snowy picture can be a symptom of the heads rotating at the wrong speed (the cylinder head motor is binding or needs lubricating or worn video heads (in this case, the bottom portion of the picture looks clear, but poor video quality, only because of extreme tracking misadjustment in a four head machine).

2 *VCR Troubleshooting Tips*

If the unused head (for a particular mode) is not shorted out, it will also read data and distort the picture. In SP mode the SP heads read the video tape first and then the EP heads contact the tape. If the EP heads are not shorted out the bottom portion of the picture is distorted. Similarly in EP mode the SP heads contact the tape before the EP heads and again if they are not shorted, the top part of the picture is affected.

Noisy Picture or Absence of Video

When you observe a noisy picture, or if there is no video at all, check the following possible causes.

- Video heads (if the problem occurs at all speeds)
- If the problem occurs only at one speed, or if a portion of the picture is clear, check the head switching circuits and preamplifier.

In general, check:

- Video head preamplifiers
- Playback/record switching transistor
- RF modulator
- Luminance (B/W picture) circuits
- Rotary transformer
- Power supply
- Chrominance (color) circuits
- 3.58MHz oscillator
- 4.2MHz conversion circuit
- 30Hz and 15Hz reference signals
- 629 kHz color subcarrier. For Beta, this is 688 kHz.

Note: It is possible for a worn (marginal) head to record a tape which plays back acceptably on another machine (although the picture may be somewhat degraded) but will exhibit noise when it plays back its own tapes. On older VCR's with fine tuning channel controls, if the problem does not affect all channels, check the thumbwheel setting. The thumbwheel may be improperly set or may be intermittent due to dust and dirt.

VCR Troubleshooting Tips **2**

Luminance problems, or a poor definition picture lacking sharpness may be related to the video preamp circuit or the noise canceller circuit.

Noise or snow may indicate problems in the tracking control circuit, or dirty heads. If the snow is coarse grained check for dirty heads. Dirty heads account for about half of the noise or snow associated with VCR problems.

A herringbone pattern on the screen when playing a known good tape may be the result of a carrier leak, check the FM demodulators and limiters. If the herringbone pattern occurs in a color picture but not in a BAV picture (turn TV color controls off to check this) check color/ luminance circuits for leakage.

A narrow band of noise at the bottom of the picture indicates that the heads are switching too soon. If you encounter this symptom, check the 30Hz control track logic pulses and 30Hz reference signal.

The function of the delay line is to replace missing data, generally caused by tape oxide defects, with data from the previous horizontal scan line. This circuit will not affect recordings. It only functions in playback. If a recorded tape has a lot of random dropouts (not a snowy picture) but plays well on a known good VCR suspect the delay line circuitry.

Flagging

If you observe flagging, that is the upper part of the picture is skewed or bent over (a wave in the top part of the picture), check:

- Tape back tension
- TV AFC circuit

TV's built before 1978 have a lower video sync tolerance and the AFC circuit may not follow the VCR playback. If this is the problem, the TV AFC circuit needs to be modified (capacitor values changed).

Some prerecorded tapes with copy protection can also cause this if the output level is reduced by going through a switching system or another VCR.

2 VCR Troubleshooting Tips

No Picture In TV Mode

When the VCR is in TV mode, and there is no picture, check:

- TV/VCR switch
- TV/VCR switching relay

Either may have dirty or worn contacts.

No Picture In Stop Mode

When the VCR is in stop mode, and there is no picture, check:

- Tuner/IF
- Video input connector (see *Figure 2-2*)
- Mode switch

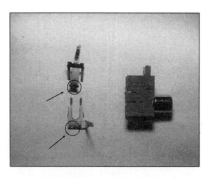

Figure 2-2. Oxidized input jacks can result in loss of audio or video signals when making recordings.

Color-related Picture Problems

Defects in VCR circuits frequently cause color-related problems in the playback picture such as absence of color, flickering of color, or other picture defects in which color is incorrect. Here are some typical symptoms accompanied by suggested areas to check.

VCR Troubleshooting Tips 2

1. No color. Check:

    - Color playback or reference circuits
    - *AFC circuit*
    - Dropout compensation circuit (if this problem occurred after a severe dropout)

2. Loss of color lock, barber pole effect. Check:

    - FC and horizontal sync pulses
    - VCO frequency

3. Flickering of color in playback. Check:

    - Automatic color control circuit
    - Video heads (especially if the problem shows up in a B/W picture as well)
    - Video head preamplifier balance

4. Color change in playback. Check:

    - Color subcarrier frequency

5. Bands of color several lines wide, other colors saturated, check:

    - Automatic phase control circuits
    - 3.58MHz oscillator frequency

Audio Problems

In addition to video problems, VCRs may also exhibit audio problems. If the unit has a hi-fi sound system, the audio problems may be related to either the linear audio or the hi-fi audio. The problem may be related to the record/playback heads, or may be caused by the circuitry. Following are some typical audio problems you may encounter along with suggestions of areas to investigate.

2 VCR Troubleshooting Tips

1. Non-hi-fi audio problem but hi-fi audio normal. Check:

    - Stationary audio/servo head
    - Audio circuit, especially record/playback relay (contacts may be dirty) if recorded tapes made on the problem machine play normally on a different VCR.

2. No sound from TV program in stop mode. Check:

    - TV/VCR relay (dirty or worn)
    - Tuner/IF - RF converter
    - Mode switch
    - Audio input connector (*Figure 2-2*)

3. Hi-fi sound problem on playback, standard audio good. Check:

    - Hi-fi audio heads (on cylinder)
    - Rotary transformer
    - Audio preamplifiers
    - Audio head switching relay

4. Hi-fi audio sounds raspy. Check:

    - Head switching relay and circuitry
    - Audio dropout detection circuitry

The audio heads in a hi-fi VHS machine are located on the cylinder along with the video heads, and are switched at a 3OHz rate. The audio heads are 120 degrees around the head cylinder from the adjacent video head, and have an azimuth angle of +30 degrees. Audio head switching is delayed slightly from the video head switching by an audio phase shifter circuit. Incorrect audio head shifting results in loss of, or noisy, hi-fi audio.

In operation, the audio signals are recorded first, then the video signals recorded over them. The different head gaps enable this depth multiplexing where the audio signals are recorded deep into the tape, with the video signals only recorded to a shallow depth. The audio heads reject video data due to their +30-degree azimuth and further video signal rejection is done by bandpass filtering of the audio head signals.

In Beta machines the FM luminance frequency is moved up a small amount so that the sidebands of audio converted to FM signals will not overlap with the sidebands of the FM luminance and down-converted chroma. Therefore, different head azimuths are not needed to keep the audio converted FM signals separate from the video FM.

Beta machines use the video heads to also record hi-fi audio using four different audio-to-FM frequencies. Filters switch in during playback to prevent audio tracks from interfering with the video information.

Audio and Video Problems

1. Noise bar in the picture; sound intermittent, absent or garbled. Check:

    - Audio/control head

2. Garbled or off-pitch audio, out of sync picture. Check:

    - Capstan

3. Picture pulsates in and out at a steady rate, no audio. Check:

    - Capstan servo lock (the absence of audio is caused by the VCR cutting off the sound because the servo is not locked)

4. Playback speed seems incorrect, picture noisy and sound pitch too high or low. Check:

    - Mode switch
    - Servo head (look for this in the audio/servo head stack)
    - Capstan servo circuit

5. Jittery video and garbled audio. Check:

    - Tape movement (*Figures 2-3* and *2-4*)

Figure 2-3. Irregular roller movement (leading to picture jitter and distorted audio) can be detected by looking at the reflections in the moving tape. If any jitter at all is noted in the reflection, the roller needs to be adjusted or replaced.

2 VCR Troubleshooting Tips

Figure 2-4. Dirty reel tension brakes can cause erratic reel movement which can result in picture jitter and audio garbling.

Tape Interchange Problems

If known-good tapes from other VCRs play poorly, or the tracking control can not be adjusted to eliminate white streaks in the picture, the tape path probably needs to be adjusted.

VCR Control Problems

1. VCR remains in, or goes into, Stop mode. Check:

    - Dew sensor circuit
    - End of tape sensors
    - Movement of reels when loading tape
    - Reel sensors - Tape slack sensor (microswitch)
    - Counter belt (*Figure 2-5*)
    - Counter circuit (and check Hall Effect sensors under reels)
    - VCR cassette sensor switch

Figure 2-5. A loose footage counter belt will cause VCR shutdown in machines without reel sensors (the system microprocessor will not receive signals through the footage counter feedback circuitry and think the take up reel is not turning).

VCR Troubleshooting Tips

2. VCR accepts cassette (front load unit) but will not load tape. Check:

- End of tape sensors
- Cylinder rotation for servo lock

3. VCR accepts cassette but then loads and unloads tape continuously (or shows these symptoms when play or record mode selected), or the VCR will not accept cassette, check:

- End of tape sensors
- Rewind sensor

If you hear the tape loading motor. Check:

- Tape loading gears (check to see if they are worn or broken)
- Cassette eject switch (they may be mechanically stuck "on" or shorted)
- Mode switch
- Cassette tray (stage) motor and belt (*Figure 2-6*)

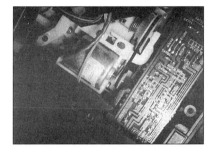

Figure 2-6. The marks on the belt pulley, and debris under the pulley indicate that this belt is dried out and needs replacement.

4. VCR goes into Stop mode and unloads tape. Check:

- Record tab switch (if this switch is intermittent it may cause these symptoms, especially in Matsushita, Panasonic, GE and Magnavox VCR's.) - The play pulley
- The play roller
- Felt clutch (in some RCA and Panasonic units, improper torque can cause this symptom.)

5. VCR does not respond to input switches. Check:

2 VCR Troubleshooting Tips

- Power supply
- State switch
- System control circuits
- Sensors (especially the end of tape sensor lamp)
- Motors

6. VCR will not go into record mode or on-timed recordings did not record. Check:

- Record tab (see below)
- Record tab Microswitch

When the problem is intermittent, it is important to verify that it was not actually caused by one cassette used intermittently. This problem occurred in one VCR because a cassette had a tab that was slightly out of tolerance and did not always activate the record tab switch. Thus the VCR would not always record in the timer mode when this cassette was being used.

7. VCR does not fast forward or rewind, or is slow in these modes (cue/review function normal). Check:

- Clutch tires (*Figure 2-8*)
- Idler assembly

In addition to mechanical problems, cheap no name type video cassettes may have poor hub lock mechanisms that do not fully disengage, making it hard, if not impossible for the VCR to turn the cassette tape reels.

Another possibility is foreign objects that are interfering with the mechanism. Inspect the interior of the VCR carefully for foreign objects (*Figure 2-7*).

Figure 2-7. VCR problem caused by a worn clutch tire, idler mechanism or objects lodged in the machine.

Figure 2-8. A common VCR problem is "eating tape" caused by tape not retracting back into the cartridge. (see Figure 2-7).

VCR Troubleshooting Tips

8. VCR eats tape because the tape is not drawn back into the cassette. Check:

- Clutch tires
- Idler assembly

The idlers are usually found between the reels on the top of the chassis, and are therefore relatively easy to service (on some models the idler can be lifted out of the VCR after a spring is removed. On other models, a chassis part may have to be removed to provide enough room to maneuver the idler out of the VCR. Some Fisher units have the idlers mounted underneath the chassis thus causing longer more complicated repairs, and some Emerson models require that a circuit board above the chassis be removed.

Rubber revitalizer is only a temporary fix for belts and clutch tires which show wear but are not at the replacement stage (with a cracked or glazed shiny surface). The most the rubber revitalizer will do is add a few months to the tire or belt and is not an alternative to the replacement of these items.

Miscellaneous Symptoms

Low power supply voltages, or spikes on the dc outputs can cause problems that appear to be sensor related. Low voltages can reduce signals into uncertain areas between definite high and low logic switching and front panel control may no longer operate, or may be erratic. Spikes can upset sensor circuitry into thinking sensor pulses are present. For example, some VCRs may shut off during a lightning storm as power line spikes cause transients that fool the VCR circuitry into thinking a turn off signal occurred. Intermittent record tab switches, especially in Matsushita, Panasonic, GE and Magnavox units may produce a variety of symptoms depending on contact resistance. Symptoms include:

- Loss of timer control
- Loss of one-touch recording

VCR going in and out of record mode but front panel indicators do not show this happening:

- Snowy picture
- Hissing audio with noises, clicks or motorboating sounds
- New video is recorded but the original audio remains (when recording over a recorded tape). This can also be caused by poor contacts in an audio input jack.

17

Chapter 3
Video Update: Setting VCR Head Switching
By the ES&T Staff

This article is based on Tech Tip 108 from Sencore. All artwork is courtesy of Sencore.

VCR technicians have a need to know how to set the head-switching signals in VCRs. Another name for this adjustment is the "PG Shifter" control. This article, based on Sencore Tech Tip 108, explains this adjustment in detail.

We will start by explaining how the head switching adjustment affects VCR performance. We will then explain two ways to adjust the circuits using the oscilloscope. The first method is based on using the scope in a conventional manner, manually counting sync pulses. The second method assumes that the oscilloscope available is one of the newer more sophisticated scopes with a delta time function. The delta time method can also be used for any other VCR adjustment that needs a time delay between two signals, such as the tracking-fix (sometimes called tracking preset) adjustment and the timing of the hi-fi heads in VHS tape decks.

Why Head Switching Needs Adjustment

Before we explore how to set the head-switching signal, lets consider what it does. Every VCR uses a pair of video heads when playing a tape at normal speed. Even decks with 3, 4 or 5 video heads use the heads two at a time. (See *Figure 3-1*).

A 30Hz square wave from the servo circuits controls an electronic switch at the head amplifier output. The switch selects the amplifier for the head which is in contact with the tape and turns off the channel for the head which is on the opposite side of the drum. If the second head was not turned off it would add noise to the playback signal.

3 Video Update: Setting VCR Head-Switching

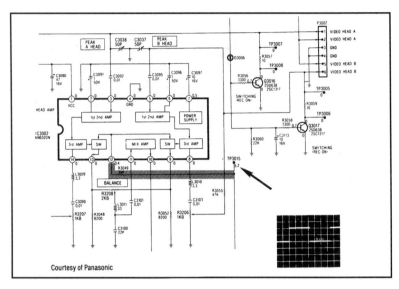

Figure 3-1. *The head switcher uses a 30 Hz square wave from the servo circuits to turn off the amplifiers of the head which is not contacting the tape.*

Noise appears in the video signal when the switching takes place. (See *Figure 3-2*). You can see this noise by adjusting the vertical hold control to display the sync interval on a TV connected to the VCR. The switching noise is a horizontal tear in the picture a few horizontal lines above the black sync bar.

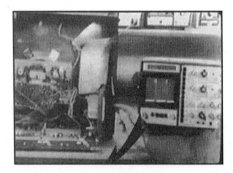

Figure 3-2. *Noise appears in the picture at the point where the VCR switches from one video head to the other. The switching adjustment keeps the noise close to the bottom of the screen, so that it is not annoying.*

The head switching circuits change the timing of the switching signal with reference to vertical sync. Switching should take place a few lines before vertical blanking to place the noise in the bottom 3 lines of the picture. Since most TVs are overscanned (the vertical deflection is slightly larger than the CRT screen), switching is invisible, because it happens while the electron beam is below the screen. If the circuits switch too early, the noise moves up into the visible part of the picture. If the circuits switch too late, the noise occurs during the sync pulse, causing poor vertical stability.

Now that you understand how the adjustment affects the circuits, you should have a better understanding of why the timing must be correct. This understanding should also help understand the alignment procedures. Now let's see how to adjust the pulse timing. We will start with the conventional oscilloscope method.

Adjusting Head Switching By Counting Pulses

The first thing you need to do is locate the test points and the controls that affect the head switching. The service literature for the VCR you are servicing is the best source of this information. The service literature also tells you how many adjustments the VCR contains. (See *Figure 3-3*.).

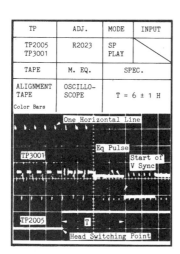

Figure 3-3. This is a typical manufacturer's head switching instruction. Use the service literature to determine the test points and adjustments to use for the adjustment.

Depending on the VCR, it may have one, two, or three adjustments. Most 2-head VCRs have only one control. VCRs with four (or more) video heads may have two playback adjustments. If so, you will need a test tape recorded at the fastest tape speed (SP or Beta 1) to adjust one control, and a tape recorded at the slowest speed (EP or Beta 111) to adjust the other. Some early VCRs also have a third adjustment in the recording circuits.

3 Video Update: Setting VCR Head-Switching

The instructions will usually tell you to adjust the control until the switching square wave is 6.5 horizontal lines ahead of vertical sync. If you are counting pulses to make this adjustment, remember that you must count every other pulse through the blanking interval if your test tape has interlaced sync. This happens because the vertical blanking pulses contain equalizing pulses at twice the rate of the horizontal sync pulses.

If your tape has non-interlaced sync, it may not contain equalizing pulses, so you must count *every* pulse. You can avoid the question of whether to skip pulses by remembering that the blanking interval is always three horizontal lines wide. Count 3.5 horizontal lines from the start of blanking instead of 6.5 lines from sync. This lets you use the same procedure, whether or not your signal contains equalizing pulses.

Using a Dual-trace Scope With Delta Time Capability

If your scope has two input channels and a delta time capability, you can use it to help you adjust the head switching, and avoid counting of pulses. Refer to the VCR manufacturer's service literature to find the needed test points and adjustment locations. Then, use the following procedures to make each head switching adjustment using the oscilloscope screen. *Figure 3-4* shows the sequences of steps for this procedure for one manufacturer's waveform analyzer.

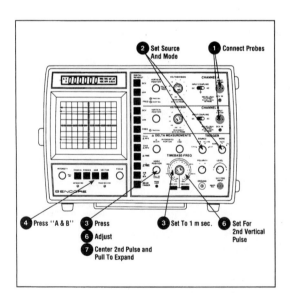

Figure 3-4. This drawing shows the sequence of steps that a technician would use to display the head-switching signal on the Sencore Waveform Analyzer.

Video Update: Setting VCR Head-Switching   **3**

1. Connect the Channel A probe to the VCR video output and the Channel B probe to the test point with the head-switching square wave signal.

2. Set the scope's TRIGGER SOURCE switch to the "Channel B" position (to trigger from the square wave) and the TRIGGER MODE switch to "AUTO." The TRIGGER POLARITY switch lets you select the rising or falling transition, depending on which one you want to use.

3. Set the TIMEBASE-FREQ switch to the 1 msec position (check the HORIZ POSITION control to confirm that it's in the correct position for a non-expanded trace).

4. Press the A&B (dual trace) selector button and adjust the inputs and triggering circuit until the two traces are locked in on the scope face.

5. Place the VCR into the record or playback mode, depending on the manufacturer's alignment instructions.

6. With the trace positioned to start at the left side of the CRT, adjust the horizontal vernier control (the small knob in the center of the TIMEBASE-FREQ control) until you see two vertical sync pulses on the channel A trace - one at the left edge and the second one near the right edge of the screen. (Channel B should show a square wave transition near the second sync pulse.)

7. Adjust the HORIZONTAL POSITION control until the right hand vertical sync pulse (and square wave transition) is in the center of the screen. Set the HORIZ POSITION control to the correct position to expand the waveforms by ten times.

8. Carefully watch the trace as you adjust the control. Start by adjusting the timing until the square wave just touches the vertical sync pulse. Then move the transition to the beginning of vertical blanking. Finally, move the transitions 3.5 horizontal lines before blanking (which is the same as 6.5 lines ahead of vertical sync).

9. Some people prefer to add channel A to channel B by manipulating the appropriate controls. This makes it easier to compare the timing of the two signals. When added, the square wave causes a step to appear in the video waveform.

Adjust the head-switching control until the step is 3.5 horizontal lines ahead of the vertical blanking (see *Figure 3-5*).

3 Video Update: Setting VCR Head-Switching

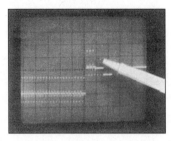

Figure 3-5. Setting the oscilloscope up so that Channel A and Channel B are added together results in a display that shows a jog at the point where the heads switch.

Using Delta Time to Adjust Head Switching

If your oscilloscope has a delta time function, you can use it to eliminate the need to count pulses. You preset the delta begin and delta end controls until the digital readout shows the correct time, and then adjust the head switch control until the sync pulse touches the highlighted area of the waveform. To use the delta time function, you need to know how many microseconds to leave between the square wave and the sync pulse. Simply multiply the lines specified by the time for one horizontal line: 63.5 μsec. Your servicing instructions may use one of three delays: 6, 6.5 or 7 horizontal lines. *Figure 3-6* shows the calculated values for each delay. Lock the waveforms onto the CRT by following the previous steps 1 through 7. Then, choose the delta time function and set the interval so that the interval begins just to the left of the sync pulse, and ends to the right of the transition, such that the readout shows the correct time interval. Finally, adjust the head-switch control until the vertical sync pulse just touches the end of the time interval.

LINES	MICROSECONDS
6	381
6.5	413
7	444

Figure 3-6. The number of microseconds that correspond to typical head-switching specifications.

To use the delta time feature to set head switching: (See *Figure 3-7.*)

1. Follow steps 1 through 7 from above to display the two signals on the scope face.

2. Make whatever adjustments are necessary on the oscilloscope so that you can easily see the area of waveform that is of interest.

3. Select as the beginning of the time interval the square wave transition in channel B.

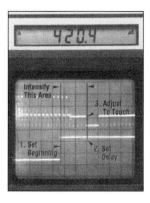

Figure 3-7. Use the delta time function, if your oscilloscope has this feature, to preset the time needed between signal, and then adjust the circuits until the signals touch the intensified area of the waveform.

4. Select the end of the desired interval such that the digital or on-screen readout shows the correct time for the waveform (for example, 413 μsec for a 6.5 line delay. Don't be too fussy in this setting since the circuits only need to be adjusted within 30 μsec of the ideal amount.

5. Adjust the VCR head-switch control until the beginning of the vertical sync pulse just touches the end of the selected time interval.

You can use a similar procedure any time you need to set a time delay between the signals at two test points.

Chapter 4
Movable Heads Drive Circuitry
By the ES&T Staff

This article is adapted from the Expander, a monthly publication that Mitsubishi Electronics America, Inc.-Technical Services Division publishes for its authorized service centers.

In the speed search mode, the tape in a VCR travels at a much faster speed than it does during the normal play mode, so the video heads describe a path across. The tape that is much closer to vertical than diagonal. Therefore, instead of tracing out the recorded stripe of information, the head traverses more than one video track.

The result is noise bars in the picture when the VCR is in the speed search mode. This phenomenon was described in detail in "Video Comer" in the August 1991 issue.

Addition of a set of video heads to the head drum that could move up and down relative to the circular face of the head drum, as. The drum rotates and the tape is pulled past the head drum eliminates these noise bars. This article looks at the movable heads drive circuitry in more detail, covering the generation and control of the actual drive signals.

Block Diagram Analysis

A basic block diagram of the movable heads drive circuitry is illustrated in *Figure 4-1*. When a movable head (MV) mode is activated, either speed search or all jog/shuttle modes except high speed search, the channel 1 and channel 2 drive generators output a sawtooth signal to drive the movable heads actuator coils. This is only a rough signal. The amplitude and average dc level must be adjusted to produce a noise free picture.

The control of the drive signals starts at the output of the MV heads. When an MV mode is activated, the head selection circuitry switches the signal source from the normal playback 4 x heads to the MV heads. The output of the MV

4 Movable Heads Drive Circuitry

heads is directed to the video/chroma playback circuitry (not shown in *Figure 4-1*) and to signal detection circuitry.

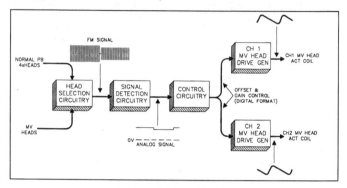

Figure 4-1. MV head drive circuit block diagram.

The signal detection circuitry outputs an analog voltage denoting the presence and amplitude of the FM signal from the heads. If no signal is present the output approaches zero volts. When a signal is present the output goes positive. The value of the positive output is directly proportional to the amplitude of the input FM signal. If there is a slight difference in the output of the individual heads, it is indicated in the output of the signal detection circuitry, as shown in *Figure 4-1*. This does not hinder the operation of the circuitry as long as the output of the signal detection circuit indicates FM signal is present.

The signal detection output is applied to the control circuitry of the VCR. This determines the required changes in amplitude and dc level of the drive signals to obtain maximum output from the heads. The control circuitry outputs the control data in a digital format which is applied to the two MV head drive generators altering the drive signals accordingly.

Head Selection and Signal Detection Circuitry

A simplified version of the head selection circuitry is shown in *Figure 4-2*. ICO1 contains the preamplifiers and head switching circuitry for the 4 x head configuration, and is the signal source in normal playback. The output of ICO1 is directed to pins 2 and 3 of analog switch IC04. During normal playback the signal from ICO1 is output at pin 15 and 4 of IC04. The signal from pin 15 is directed to the video/chroma circuitry, and from pin 4 to the envelope detector in IC2A1 (signal detection circuit). In playback, the output of the envelope detector controls the auto tracking function of the VCR.

Movable Heads Drive Circuitry 4

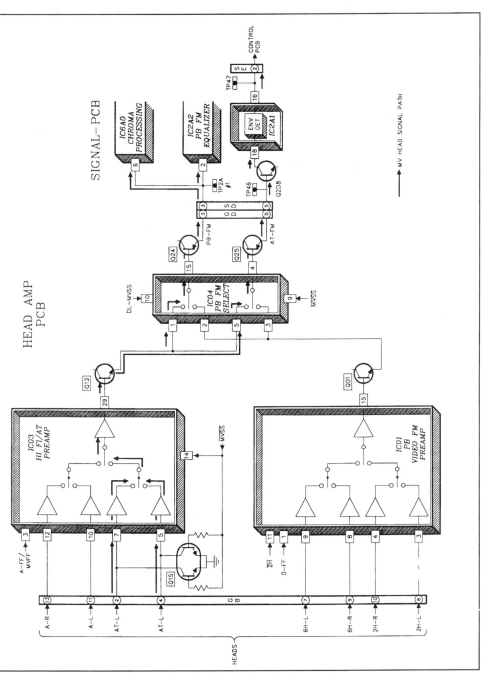

Figure 4-2. Movable heads selection circuitry.

4   Movable Heads Drive Circuitry

When an MV mode is activated the MVSS control line goes low. The MVSS line controls the lower switch in IC04 and the head selection switch in IC04. IC03 is normally the preamp for the hi-fi heads. However, when the MVSS line goes low, the MV heads are selected for the output at pin 29 of the IC. The signal from the movable heads is routed through IC04 to the envelope detector. The resulting MV head output status is directed to the control circuitry, which generates correction data for the two MV drive generators.

Approximately 1 second after the MVSS line goes low, the DL-MVSS line goes high, switching the upper analog switch in IC04 to the MV head output, directing it to the video/chroma circuitry. The slight delay between the change on the MVSS and DL-MVSS lines allows the MV drive signal to be adjusted before the MV heads output is utilized as the signal source. Due to the delay, the first picture displayed when entering an MV head mode, is from the conventional 4x heads and noise bars are momentarily present. Then it switches to the MV heads and a clean picture is produced.

MV Heads Drive Signal Generation

A simplified version of the MV drive generator circuitry is illustrated in *Figure 4-3*. Note that the output of the envelope detector takes two paths: to the V-ENV input of IC5AO, the mechanical control µPC; and to the AT-ENV input of IC401, the NE (noise elimination) µPC. During normal playback, IC5AO utilizes the output of the envelope detector to generate control signals. These are output at pin 36 in a serial format and are directed to the auto tracking circuitry in the servo circuit.

When an MV head mode is activated the two drive generator ICs, IC403 and IC404, generate rough sawtooth drive signals for the MV heads. At the same time the AT-ENV input of IC401 is enabled. Internal to IC401, the signal at the AT-ENV input is analyzed and digital correction signals are generated. Correction signals controlling the average dc devel (offset) of the drive signal are output in a serial format at pin 34 and are directed to the two drive generator ICs. Amplitudes of the drive signals are controlled by parallel data gain control lines directed to each drive generator IC. Gain control for channel 1 (IC403) is output at pins 52 through 56, and gain control for channel 2 (IC404) is output at pins 44 through 48.

Movable Heads Drive Circuitry 4

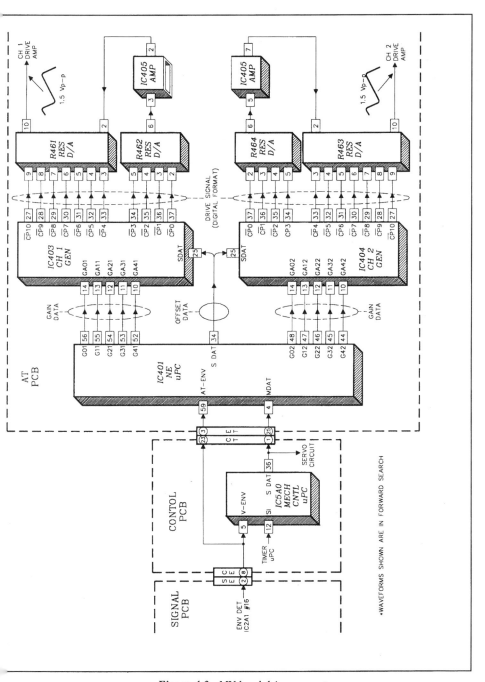

Figure 4-3. MV head drive generators.

4   Movable Heads Drive Circuitry

The outputs of IC403 and IC404 are 11 parallel digital data lines that are directed to resistive network D/A converters, which convert the digital data to the required analog sawtooth drive. R461 and R462 comprise the D/A converter for channel 1, and R463 and R464 make up the D/A converter for channel 2. The output of the D/A converters are directed to drive amplifiers for further amplification before being applied to the movable heads. The waveforms shown in Figure 3 are typical of the drive signals generated in forward speed search.

Before looking at the drive amplifiers it should be noted that IC401 is not only utilized in developing MV head drive, but is also part of the overall control circuitry. Although it may not be shown in *Figure 4-3*, it is also used to generate conventional mode operational commands such as: CT-REV (counter reverse), NOT SS (speed search), EE, STILL (conventional PB still), and SL (slow motion).

MV Drive Amplifier Circuitry

The MV drive amplifier circuitry is shown in *Figure 4-4*. Basically the circuitry is relatively straight forward. The drive signals are amplified by amplifiers in IC406 and IC407. VR402 and VR403 set the initial offset (dc level) of their respective channels. VR404 and VR405 determine the gain by adjusting the amount of negative feedback to the input of the amplifiers in IC407. Note the output from IC407 is not applied directly to the MV heads. It is directed to K401, a signal selection relay, which routes the drive signals to the MV heads actuator coils.

The relay is required since the brush contacts on the top of the drum assembly are used for two purposes: to couple the MV drive signals to the upper drum and also to supply the flying erase heads signal at the start of a recording. *Figure 4-5* illustrates the relay circuitry. Normally the relay is de-energized and the contacts are in the upper position, connecting the MV actuator coils to the outputs of IC407. The flying erase heads are basically in parallel with the MV actuator coils. When MV head drive is present, the high impedance of the 390pF capacitor in series with the flying erase heads prevents the 30Hz MV drive signal from affecting the flying erase heads.

Movable Heads Drive Circuitry 4

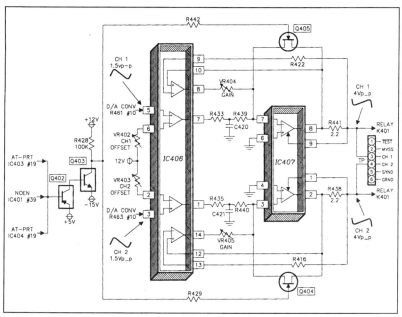

Figure 4-4. MV actuator drive and protect circuitry.

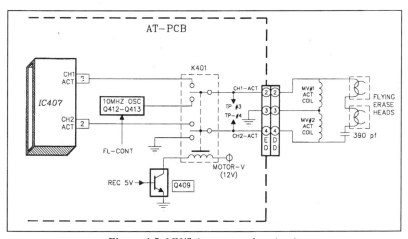

Figure 4-5. MV/flying erase relay circuit.

4 *Movable Heads Drive Circuitry*

When record is activated, the record 5V supply turns on Q409. This energizes the relay and connects the output of the 10MHz oscillator, consisting of Q412 through Q413, to the brush contacts on top of the drum assembly. The FL-CONT line activates the 10MHz oscillator for several seconds at the start of record. This erases any previously recorded signal not erased by the full erase head. When the 10MHZ flying erase signal is present, the high impedance of the actuator coils limits any current flow from the erase signal so the movable heads are not affected.

Protect Circuitry

The MV heads are protected by protect circuitry, consisting of Q402, Q403 Q404 and Q405. (Refer to *Figure 4-4*.) This circuitry is necessary to protect the MV heads actuator coils when not in an MV mode. When not in an MV mode, the output from the A/D converters drops to zero volts. This causes the output for the first drive amplifiers, pin 7 and 1, to go to approximately 4V, due to the positive offset voltage at pins 6 and 2. The -4V is inverted by IC407 and is constantly applied to the MV head actuator coils. If the VCR is allowed to remain in this condition for a long period of time the actuator coils will be damaged by the constant dc current. To prevent this from occurring, the protect circuitry disables the op amplifiers in IC407 when not in an MV head mode.

Three control lines control the protect circuit, an AT-PRT output from both IC403 and IC404, and a NOEN output from the NEmPC IC40 1. All three control lines are tied together and are connected to the base of Q402. When in an MV head mode all three lines are low, holding Q405 and Q406 OFF, through Q402 and Q403. When not in an MV head mode, all three control lines go HIGH turning on Q405 and Q406. With Q405 and Q406 conducting, the outputs of the OP-AMPS in IC407 are effectively shorted to the inverting inputs, canceling the negative dc voltage from IC406.

Troubleshooting Tips

Problems in the MV head drive circuitry only appear when an MV head mode, such as speed search or jog/shuttle is activated. In the jog/shuttle mode the severity of the symptom may vary, depending whether still, slow, playback or speed search are activated. When speed search is activated directly from playback, the symptom is usually only displayed momentarily. The VCR automati-

Movable Heads Drive Circuitry **4**

cally switches back to the normal 4 x head configuration, producing the conventional speed search picture with noise bars.

Although the movable heads are only used during special effects, it must be remembered they are still video heads and are subject to the same problems as conventional heads. If one head is defective or clogged, the resulting picture will be mostly snow with an apparent 30 Hz flicker, the same as when a conventional head is damaged or dirty. Therefore, when a MV head mode problem is encountered, clean the heads first. This may resolve the problem and eliminate any circuit troubleshooting time.

If the problem still exists after cleaning the heads, one head may still be defective. It is difficult to check the MV heads output in an MV mode with the unit malfunctioning, since the drive signal to the MV heads is constantly changing looking for an acceptable signal. We suggest using the alignment test mode.

Play a tape and connect a jumper between pins 1 and 6 of the TP on the AT-PCB. This activates the movable heads during normal playback. The output of the MV drive circuit in this mode is only an offset (dc) voltage, not a sawtooth since the head position does not have to be continually varied in playback.

To view the output of the heads, connect an oscilloscope to pin 1 of TPA on the SIGNAL-PCB and trigger the scope from the head switching signal at pin 2 of TP on the AT-PCB. If the output of one or both heads appears to be missing, it may be necessary to perform an alignment procedure. If the offset or gain are misadjusted, it may appear that no output exists from that channel's head.

If satisfactory adjustment cannot be obtained, the problem may be localized further by noting the MV drive signal offset range (channel 1 TP #3 and channel 2 TP #4) as the offset adjustment is rotated from one end to the other. The range of the offset voltage should extend from approximately +3.5V to -3.5V. If the range of the offset voltage is too low the cause is probably in the MV drive generator or MV drive amplifier circuitry. If the offset range is incorrect, it points to a problem in the MV head preamps (IC03) or the MV heads themselves.

On rare occasions the MV drive circuitry may adjust OK when pins 1 and 6 of TP are jumpered, but noise bars still exist in MV modes. Activate forward speed search and monitor the MV heads drive signals with an oscilloscope, channel 1 at pin 3 of TP and channel 2 at pin 4. The two drive signals should be approximately $4V_{pp}$. When there is a marked difference between the two signals it points to problems in the drive generator circuitry.

4   Movable Heads Drive Circuitry

If the two drive signals appear normal, check the MV head output at pin 1 of TP2A on the SIGNAL-PCB. If the output of a head is OK in the center, but falls off drastically at the start and end of the envelope, as shown in *Figure 4-6*, the actuator coil may be damaged or the head's movement mechanically restricted, which would then require an upper drum assembly replacement.

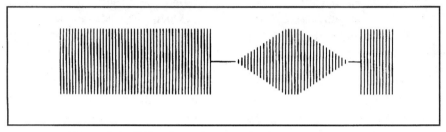

Figure 4-6. MV heads envelope (one head's movement restricted).

Chapter 5
Camcorder Electrical Adjustment
By the ES&T Staff

The digital revolution has wrought many changes in the way consumer electronics products operate. When you tune a modern radio, you no longer rotate the dial and watch the meter to find the best reception. You push a button and the radio's tuned to the exact station frequency. With CD players, you slide a disk into the player then select a track, or program the unit to play tracks in whatever order you choose, then you just sit back to listen to digitally encoded music. It's no longer necessary to carefully cue the tone arm, or do any of those other things you once had to do using vinyl disks.

Digital Adjustments

The introduction of digital circuitry has, in some ways, made things both more difficult and easier for service technicians. For example, while the microminiaturization of components used in camcorders has made them difficult to see and even more difficult to handle, the digital nature of those components has made it possible to make adjustments to the camera's operation without ever taking the cover off, or without ever turning a single wiper of a single pot. By changing the data stored in an integrated circuit in the camcorder, it's possible to adjust such characteristics as black level, sync level, burst level, and autofocus level.

Making Adjustments

Modem camcorders are controlled by microprocessors. Microprocessors are really pretty much microcomputers that are capable of performing only a limited set of functions. In older electronics products, when the voltages that control the functions of the product vary as the product ages, they are adjusted by changing the value of an adjustment potentiometer. In modem products, such

5 Camcorder Electrical Adjustment

as a camcorder, the voltages that control those functions are derived and adjusted digitally. The microprocessor establishes the value of those voltages based on data stored in a memory module. The value that is output from the microprocessor in digital form is converted by a digital to analog (D/A, or D-to-A) converter into those voltages (*Figure 5-1*).

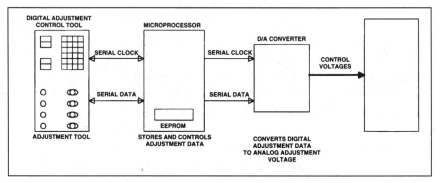

Figure 5-1. *The digital adjustment tool allows technicians to adjust the camcorder's characteristics without even removing the cover from the unit.*

As the characteristics of the components being controlled by those voltages change with age (let's say that the recorded picture on the tapes recorded by the camcorder are beginning to look smeared during playback), the technician needs to adjust the voltage that controls that characteristic. Since the value of that control voltage depends on a set of information stored in a memory module, he needs to update that information to adjust for the new condition. Camcorder manufacturers provide technicians with ways to do that.

The EEPROM

The data that the microprocessor uses to establish the control voltages is stored in an EEPROM (electrically erasable programmable read only memory). In the system shown in *Figure 5-1*, the EEPROM is actually in the same package as the microprocessor. As the term EEPROM implies, this device is a ROM, or read only memory. That's just a little misleading here, as the data within the ROM can actually be changed. However, for all intents and purposes, an EEPROM is a ROM during normal system operation. That is, until a technician deliberately uses some kind of device to reprogram the EEPROM, the information is permanent; unlike the information stored in RAM, which may be changed many times during operation of the product, and which disappears once the system is turned off.

Camcorder Electrical Adjustment **5**

Two Ways To Make Adjustments

Manufacturers provide technicians with two methods of making adjustments to the camcorder by changing the data in the EEPROM; a computer assisted adjustment system, and a digital adjustment tool. The computer assisted system consists of an interface box, a cable to connect the box to the computer, a cable to connect the box to the camcorder, and the software, on floppy disk.

The digital adjustment tool connects directly to the camcorder via a special connector, and adjustments are made by pressing the appropriate buttons on the front panel of the instrument.

Other Instruments Required

While the use of digital data for adjustment of camcorder function eliminates mechanical potentiometers and the need to get inside the camcorder to make adjustments, it does not eliminate the requirement for the other tools and test equipment familiar to consumer electronic servicing technicians. The camcorder is a complex device that requires a full complement of test equipment. According to Thomson Consumer Electronics service data, in addition to the digital adjustment tools, also necessary for proper adjustment are: oscilloscope, DVM, frequency counter, vectorscope, light meter, tripod, color monitor, lighting, charts, and a few other items.

Hexadecimal Representation

The data stored in the EEPROM is in digital form. Each individual data location consists of eight bits. One unit of data could be represented by a series of eight individual binary bits. Using binary data to represent the data would be cumbersome, and could lead to errors, so the camcorder electrical adjustment system uses hexadecimal numbers instead. Assume that the data stored at position 0,0 in the EEPROM is 4E. Each of those hexadecimal digits represents four binary digits. The hexadecimal number 4 is represented as 0100 in binary (it's equivalent to 4 in decimal). The hexadecimal number E is represented as 1110 in binary (it's equivalent to 14 in decimal). Thus if you could actually see into the EEPROM, you would see that the data stored in the eight-bit data word in position 0,0 of the EEPROM is 01001110.

5   Camcorder Electrical Adjustment

More To Come

This article merely touches the surface of electrical camcorder adjustment in order to introduce some of the concepts of electrical adjustment. In future issues, we'll consider some of these steps in greater detail.

Chapter 6
The Gentle Art of Camcorder Repair
By T.V. Kappel

Of all the high technology consumer electronic devices on the market today, the highly reliable camcorder is the bear to repair.

These little hand-held cameras are dropped, sat on, kicked, tripped and knocked over on a tripod, and, occasionally, treated little better than a hockey puck. Then, of course, they're brought in to you for repair. The symptom? "Why, it just quit working." Then you're asked, "Do you give free estimates?" Sometimes all this leads you to feel that the gentle art of karate should be used to service them.

Evaluating the Damage

The first step in evaluating and possibly repairing these complicated units is a little like medical triage. You need to determine as quickly as possible if the patient is only slightly hurt or poised at the door of the graveyard of video cameras.

The slightly hurt ones can be rushed into the operating room. The others you cover either with a, "Not worth cost of repair," black cloth estimate, or a major medical estimate that approaches the cost of a vital organ transplant. In some of these latter cases, that's exactly what you'll be doing: transplanting complete boards for ones that have been brutally cracked and even smashed. Your fingers will be crossed the whole time hoping that the operation will save the patient. They won't pay you if it doesn't.

All kidding aside, repairing camcorders is tough. For the initial evaluation, it isn't even necessary to put the best service technician on the job. You actually need your best and fastest evaluator to take the first look. This is the triage part of the initial repair.

6 *The Gentle Art of Camcorder Repair*

By the way, in spite of all this gloom and doom, these devices can be serviced profitably. This will undoubtedly help your bottom line and your reputation immensely.

The Case Tells a Tale

To quickly estimate a repair, the first step is not to power up the camera and see what the symptom is, but to closely examine the outside case of the camera itself. These new lightweight plastic covers are very forgiving to even sledgehammer blows to them. They bounce in, smash the delicate electronic boards underneath and bounce back out with little or no sign on the outside. The operative words here are, little or no. Carefully examine the outside case of the camcorder for tiny fractures, telltale scratches, and dents on the lenses. Sometimes the customer will make efforts to clean and hide these as much as possible: not to hide these damage indicators from the service technician, but from the perpetrator's parent or spouse.

If you suspect that the unit has been seriously damaged, open it and examine it carefully before you apply power. By doing so, you may prevent further damage of the smoke and fire kind with a physical examination first. If you see no sign of serious abuse, then power it up and observe for symptoms.

Damage Behind the Lens

There was one unit I examined that was drop-kicked when the neck strap came loose on one end of the clip and the camera fell nose first to the ground. The patient was in a complete coma with little or no vital signs. There was a small dent on the lens extension in front of the glass, but little sign of other damage.

When I carefully opened the case, after taking out hundreds of tiny screws, I found small printed circuit boards with surface mounted resistors, capacitors, and discrete devices. Also, little bits of gray plastic fell out onto the workbench.

If you plan to service camcorders, be sure that you have a good supply of small hand tools and screwdrivers. Standard sizes of screwdrivers are too large. Precision tools are available through many catalogs or at local hobby shops.

The Gentle Art of Camcorder Repair **6**

I've found that one of the most useful tools to have around for camcorder servicing is a hand-held, variable speed, hobby drilling and grinding rotary tool, which I've often used to grind and drill out a headless screw or create a replacement body part. This tool is worth its weight in platinum when necessary.

Back to the Operating Room

In this particular case, the camera had been dropped on its lens. The lens was driven backward, breaking the plastic mounts holding it in place and shearing off a surface mounted capacitor and two resistors positioned on a board behind the lens itself.

I obtained replacements for the surface mounted components and installed them. This required a good variable temperature soldering or desoldering tool for this delicate part of the operation. Here again, the standard iron and soldering tips would be much too large for the job. Manufacturers do make good equipment to help you with this kind of work, but it is expensive. The last one I bought was in the five hundred dollar price range with a tip assortment that helps unsolder surface mounted IC's.

After I replaced the components, I carefully examined the board for cracks, broken or pulled loose traces. Next I checked the resistances of the circuit paths of the parts that I had replaced using the ohms function of a DMM. It all seemed okay.

The hardest part of this repair was in reattaching the lens to the broken mount. I debated on ordering a new case side. This is one of those parts that must be ordered directly from the manufacturer and may come on a slow boat from the orient. Moreover, replacing this part would have added to the cost of the repair. It was a judgment call as to the extent of the damage to the case and whether it was repairable or not.

In this case, the screw mounts were intact. The screws had, however, been forcibly torn out of the plastic mounts, stripping the threading, and leaving gray plastic dust and particles to fall on my bench.

I used a slightly larger screw than the original, and applied a dab of epoxy to cover the heads and bond them to the plastic. I crossed my fingers and hoped the owner could manage to be a little more careful in the future and this unit would not come back again. Getting those screws out in the future, should it be necessary, would be a job for the high-speed, hand-held drill and grinder tool.

The electronics, mechanics and optics checked out after I fired it up and ran it through a series of tests. Everything appeared to be operating properly, so I spent the hour of a thousand screws reassembling the camcorder, filled out the paperwork, and notified the customer that it was ready. Then, with a heavy sigh, I turned to the next one.

Actually, considering the thousands of cameras out there, and the complexity of their mechanics and electronics and the abuse they are subjected to, they are extremely reliable. I'm sure the percentage of camcorders that come in for repair are a very small fraction of the hordes out there in the hands of potential abusers.

Chapter 7
VCR Servicing: Taking the Mystery Out of Video Head Replacement
By Philip M. Zorian

VCR video heads are fragile and prone to failure. Videotape oxides and tape debris can easily clog up the infinitesimal video head gap (only 38 microns wide), resulting in either a total loss of picture or distortion. Cleaning the video heads will usually remedy this problem. Eventually, however, the video heads either wear out or become damaged to the point where a cleaning has no effect, and the only remedy is a video head replacement.

Video heads (*Figure 7-1*) are small electromagnets that are mounted with great precision onto the upper video drum. Whenever video heads require replacement, the entire upper drum must be replaced. This is not a difficult procedure, but properly identifying a defect in the video heads can be a challenge. The consequence of installing new video heads with no resultant improvement in picture quality is a loss of both time and money.

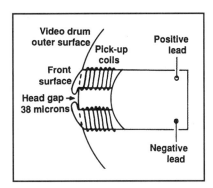

Figure 7-1. Video heads are small electromagnets that are mounted with great precision onto the upper video drum.

7 VCR Servicing: Taking the Mystery Out of Video Head Replacement

The first part of this article explains how to identify defective video heads, taking most of the risk out of this repair. The second part will guide you through the replacement of the video heads.

The Number of Video Heads

Some VCR models have two video heads while some have four or six. A common misconception is that more video heads will deliver better picture quality. The fact is, there are never more than two heads being used during the normal playback of video. The other video heads are used to improve picture quality during pause, fast forward search, special effects such as slow motion, and slower tape speeds. It is important to understand this when identifying video head defects. Since the audio heads are separate from the video heads on most models, you will typically find that, although the video is defective, the audio is normal. On VCRs with hi-fi audio, however, the audio is recorded and played back using the same heads that record and play back the video.

Diagnosing the Problem

When the video image quality is defective or when the video is completely absent but the audio is normal, suspect the video heads. The following symptoms are typical of defective video heads:

- band of static at top or bottom of screen,
- white trailing horizontal lines off right side of image,
- snowy picture,
- loss of detail and out of focus,
- partial picture,
- total picture loss,
- excessive drop-out,
- unstable image,
- one third of an image

Distorted video with distinct horizontal lines present in the white or high frequency parts of the picture are indicative of excessive head wear, since these areas are the hardest for the video head to reproduce. Total picture loss is a difficult symptom to diagnose, since it offers very little video information. The newer VCRs will display a blue screen instead of static if the video heads are

VCR Servicing: Taking the Mystery Out of Video Head Replacement   **7**

not working properly. If there is an image on the bottom third of the screen only, one of the video heads is most likely defective or clogged.

Visual Inspection

When you suspect that the heads of a VCR may be defective, start by opening the cover and spinning the video drum with your hand to be sure it spins freely. Check the two Philips screws that hold the head drum down for tightness and be sure the video drum is firmly in place. Look for any obvious damage to the drum itself. Also check the guideposts since they can occasionally break off. Make sure the tension band is in place around the base of the supply reel. Are both reels able to spin freely?

Insert a tape and press play to verify that the videotape is actually making contact with the video drum. Check the belts for slippage and the pinch roller for excessive wear. Belts are often used to drive the capstan, and a loose belt here will result in erratic tape speed. All belts are checked by holding one pulley and gently trying to turn the other to see if it slips. Use a small (20X) magnifying glass under good light to determine if the video heads are physically cracked or chipped. Get a close look at a good video head to know what you are looking for.

Cleaning the Video Heads

Clean the video heads. If this has no effect on the symptom, you can rule out the possibility that the heads are simply clogged. If this clears up the problem completely, your work is done. If you observe some slight improvement in the quality of the image, clean the heads a few more times using a solvent that is approved by the manufacturer for the purpose. You are dealing with video heads that are excessively clogged. Clean the tracking head as well.

Check the Pause Function and the Outputs

Check the pause function for information. This is helpful when dealing with a total loss of picture. A four or six head VCR may produce a picture in pause mode revealing two important things: the playback heads are defective, but all

7 *VCR Servicing: Taking the Mystery Out of Video Head Replacement*

other VCR circuits are good. This will not happen with a two head VCR, since the same heads are used for both the playback and pause functions. VCRs output video in two different ways: RF Out and Video Out. Verify that the same symptom is occurring on both video outputs. A clean video image from either one of these outputs will rule out a defect in the video heads. If you observe the exact same symptom on both outputs, the video heads are still suspect.

Check for Connection Problems

Check for a cold solder joint or loose connection on the top of the video drum. At this point you should check the video heads for continuity by removing one of the leads. An open circuit is enough evidence to indicate a defective head. There is a video head tester available on the market for about $50.00. But I have found that the test results from this tool are not conclusive, and the tester is unable to test all video heads on all VCR models.

Start by checking the simple things first. You are trying to identify, beyond a reasonable doubt, a defect in the video heads. The strategy is to rule out all other possibilities that might cause similar symptoms. At this point there are four more things to check: tape speed, tape tension, guidepost alignment and speed of rotation of the video drum.

Tape Speed

Unless the videotape is moving past the video heads at a constant rate of 131 inches per second, the video heads cannot track properly, and the video quality will suffer accordingly. Since the quality of the Audio is also affected by the tape speed, use the Audio as a way of detecting this problem by listening to a videotape with a tone recorded at 1,000Hz.

The following defects in the speed of the tape are easy to detect: too fast, too slow, wow and flutter. Be certain that you know what this tone should sound like by first playing it on a good VCR.

If you don't have a videotape with a tone recorded on it, use a tape with music you are familiar with. The ideal sound for this purpose is a single instrument or a voice. If you are able to detect a problem with the tape speed, then incorrect tape speed is the most probable cause of the video problem, and the video heads should not be replaced.

Tape Tension

Tape tension that is too low will result in a total picture loss. If the tape tension is too high, you will see symptoms that look similar to those caused by defective video heads. The first mechanism the videotape encounters as it leaves the videocassette is the tape tension guide. While monitoring the video with a known good tape, move this tension guide very slowly to the left, (to tighten) no more than 1/4 inch. Then move it slowly to the right (to loosen), no more than 1/4 inch. If you observe an improvement in the image, improper tape tension may be the cause of the video problem. This should be done on a good VCR to get a sense of the effect that tape tension can have on a VCR's ability to faithfully reproduce a clean and stable picture on the screen.

Guidepost Height

When the videotape is fully loaded, it is wrapped around 3/4 of the video drum and held in place by the guideposts. They must hold the videotape at a precise height as it moves around the drum. If the height is out of alignment, the picture quality will suffer.

A typical symptom of incorrect guidepost height is a band of distortion at either the top or bottom of the screen that by adjusting the tracking control you are unable to eliminate. Check the condition of the guideposts with the VCR turned off by gently turning the very top of the post with your fingers or a special adjustment tool. They must be tight.

If the guideposts are loose and easy to turn, the tiny set screw at the base has loosened, and the guidepost is probably in need of adjustment. You can verify this by gently moving the videotape up or down about 1/16 inch while in play mode. This must be done very carefully while monitoring the video. If the symptom clears up you can once again rule out the video heads as the cause of the problem.

7   VCR Servicing: Taking the Mystery Out of Video Head Replacement

Speed of Rotation of the Video Drum

The video drum must spin at a precise speed of 60Hz. There are two ways to verify that the video head drum speed is correct. One way is to carefully observe the rotating drum in play mode under a fluorescent lamp, this light creates a strobe affect at a similar rate to the spinning head. The video head should appear stationary or move very slowly in a counter clockwise direction.

Another way to verify that the video head drum speed is correct is to lightly touch the top of the spinning head on a smooth area, the head should attempt to overcome the resistance and react by increasing its speed.

Replacing the Video Heads

Once you have determined that the video heads are defective, the next step is to replace them. You can order a replacement upper video drum from most VCR parts vendors; they range in price from $25.00 for a two head, and up to $80.00 for a six head. Since the video heads are mounted to the video drum with great precision, the entire drum must be replaced. Do not attempt to replace the heads until you have received the new one. Always verify that it is the correct replacement.

To replace the video head drum:

1. Remove the anti-static bar; one Philips screw.
2. Remove the two Philips screws on the top of the drum.

There are typically two types of drums. Here is the way to handle each type .

- Type 1: Desolder the wires that protrude up from below the video drum and are soldered to the top surface. Once these wires are free the drum will lift off.

- Type 2: Observe the printed circuit board on the top of the drum. This type has solid leads that protrude straight through from underneath the drum. Using a solder sucker or a desoldering braid to desolder these leads as they are attached to the circuit board. Once these leads are free, the drum will lift right off.

VCR Servicing: Taking the Mystery Out of Video Head Replacement 7

If you find the drum is stuck, you will need a special tool called a video head puller. If the drum does not lift off easily, use a puller to remove it, since you should never use excessive force when it comes to the video head drum. Video head pullers are available from most VCR parts vendors for about $15.00.

Installing the Replacement Drum

Installing the replacement drum is the exact opposite of the procedure you used to remove the defective drum. Be certain to seat the new drum correctly and check for wobble by spinning it. The new drum must be put on the exact same way as the old one. Polarity matters; it can only go on one way, so pay close attention to the color of the wires and the colors on the top of the drum. Also, be extremely careful not to grab the new drum by the video heads, and avoid getting skin oils onto the drum; wearing plastic gloves is advisable.

Conclusion

Video head replacement is a repair that offers the satisfaction of instant results. Once the new drum is installed, the symptoms usually disappear and the VCR is restored back to working condition. And even considering the low cost of VCRs in today's market, VCR head drum replacement is still an economical repair.

Chapter 8
Mechanical Problems in the Sanyo VHR9300
By Steve Babbert

The unit that's most likely to come across my bench at any given time is the Sanyo VCR, VHR9300. I don't know if this is because it is prone to failure or if it is because a large number of them were sold. Whatever the reason, I think most service centers can expect to see a lot of them in the future.

As in many newer designs, this unit uses gears instead of the rubber "idler tire" to drive the feed and take-up reels. There are two belts associated with the main drive and loading mechanisms which seem to hold up as well as any. There are, however, two mechanical problems which account for a large number of failures. These problems may also develop in the VHR9200, which uses the same tape mechanism.

The Broken Front-load Gear

There is a small gear in the tape loading mechanism that transfers motion from the main loading gear to the rack that controls the forward and reverse movement of the stage. This gear is fragile, and I have found many of them with a missing tooth. This fragility might be a design flaw, or it might be intended to protect more critical parts of the mechanism in the event that someone applies too much force during the loading of a tape cassette.

When this gear fails, in addition to the obvious symptoms (no loading or unloading) there can be other symptoms that might lead the technician in the wrong direction. The System Control portion of the VCR monitors various sensors to determine the position of the tape during loading and unloading. When the gears go out of sync the Syscon may get mixed messages which can cause symptoms ranging from gear-grinding to shutdown.

8 Mechanical Problems in the Sanyo VHR9300

Stuck Tape

In one VHR9300 the tape became stuck halfway through the loading process. Power would come on but there was no response when any of the front panel buttons were pressed. Knowing this model, the first thing I checked was the infamous 16-tooth gear. As I suspected, a tooth was missing.

The first time I encountered this problem I didn't have service literature so I used a known-good unit as a reference. From the known-good unit, I then made a diagram of the position of the gears in the fully loaded and unloaded position. I found that gear replacement is much easier in the loaded position and whenever the VCR is in this position, the only gear that needs to be in a specific position during reassembly is the main loading gear.

The main circuit board must be removed to uncover the loading motor and gear assembly (*Figure 8-1*). You'll have to remove three screws from the upper bracket to gain access to the gears. The screw holding the dew sensor to the bracket does not need to be removed.

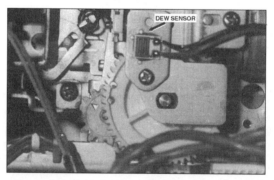

Figure 8-1. In the Sanyo VHR9300, in order to gain access to the loading motor and gear assembly, remove the upper bracket.

When the bracket is removed the worm gear will separate from the motor pulley but will remain attached to the bracket. The bracket can then be laid to the side without unplugging the dew sensor. If possible, rotate the main gear by hand to the position shown in *Figure 8-2* before removing. This is the normal position of the gear when a tape is fully loaded. Two screws which hold the small circuit board beside the gear assembly must be loosened, otherwise the connector will prevent removal of the main gear. At this point the broken gear can be removed from the VCR (*Figure 8-3*).

Mechanical Problems in the Sanyo VHR9300   8

Figure 8-2. If possible, rotate the main gear by hand to this position before removing it. This is the normal position of the gear when a tape is fully loaded into the VCR.

If the mechanism at the base of the main gear pin is in the position shown in *Figure 8-3* it will need to be moved to the position shown in *Figure 8-4*. This is somewhat difficult because it is spring loaded and tends to resist any attempt to get it into position. In most cases it will be in the right position if the main gear was removed correctly.

Figure 8-3. The cause of the failure of this VCR to load completely was this broken gear. If the mechanism at the base of the main gear pin is in this position, it will have to be moved to the position shown in Figure 8-4 in order to remove the broken gear.

Figure 8-4. The mechanism at the base of the main gear pin must be in this position in order to remove the broken loading gear.

8 *Mechanical Problems in the Sanyo VHR9300*

Once the broken gear is replaced, the rack should be pushed fully forward before installing the main gear. The small single hole in the gear should be next to the first pin on the connector Notice that the section of the gear without teeth is where the worm gear will fit. Notice also the position of the rack. At this point the housing holding the worm gear and dew sensor can be reinstalled.

Improperly installed gears can cause a variety of symptoms. One common symptom is the illumination of the "cassette in" symbol on the vacuum fluorescent display when there is no tape in the machine.

Fast Forward/Rewind Problems

Another problem that I've found in several of these chassis is a sluggish fast forward and rewind function, or the complete absence of them. The first time I encountered this problem I suspected the usual wornidler tire. However, after removing the cover and finding that gears were used instead of an idler wheel I realized that I could be in for a difficult time.

Visual inspection revealed no problems. All of the gears seemed to turn freely and mesh properly. I installed a test jig which allowed me to see if something was slipping or binding while the machine was running. I checked the take-up and feed reels by hand while the unit was both the fast forward and rewind modes. Both reels seemed to have insufficient torque. Torque was fine in the play and high speed search modes (in both forward and reverse directions.)

Close visual inspection revealed black particles around the area where the brakes came in contact with the drive gears. The brake pads had a black covering over the felt pads. I assumed that this was where the particles came from. I wondered if this deterioration was normal *(Figure 8-5)*

I ran the machine through the various cycles once again while focusing my attention on the brakes. The brakes did not lift off of the gears in either the fast forward or the rewind mode. I felt certain that they should be releasing. I checked a known good machine which showed that they should release in all modes, only making momentary contact when stopping the mechanism.

After repeated cycling of the machine between rewind and stop I noticed that the brakes would release briefly at the beginning of the rewind cycle. As soon as the machine settled into the rewind mode the brakes would snap back as if a latching mechanism was failing.

Mechanical Problems in the Sanyo VHR9300   8

Figure 8-5. Black particles around the brakes made me suspect a problem in the area.

Once again I turned my attention to the known good unit. Looking at the bottom of the unit I could see the linkage that controlled the movement of the brakes. However, it looked as though it was associated with a number of other mechanical functions. I couldn't see anything that resembled a latch. Portions of this linkage were obscured by a layered portion of the mechanism. I then returned to the top side of the machine to a point above the place where the linkage became buried underneath. Finally I found a latch that looked as though it was associated with the brakes. Returning to the defective machine I verified that the latch was indeed letting go, causing the brakes to snap back to the engaged position.

The latching mechanism is very peculiar. The hook catches on a pin that is concentric with and part of the load motor pulley (*Figure 8-6*). There is a spring that keeps a downward pressure on the latch pulling it towards the pulley. When the machine stops, the pulley reverses direction causing the hook to roll up and off of the pin, releasing the brakes.

Since I could see no signs of excessive wear on the latch itself I concluded that the spring had lost tension. Replacement with a slightly tighter spring will solve the problem, but using a spring that is too tight may prevent unlatching altogether. Once the spring is replaced and the machine is cycled, the brake will be in the position shown in *Figure 8-7*.

8 Mechanical Problems in the Sanyo VHR9300

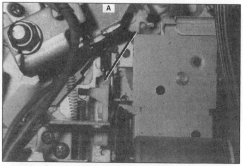

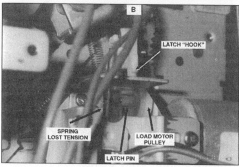

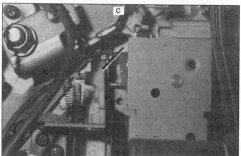

Figure 8-6. This latching mechanism, shown in its latched position (a), holds the brake pads away from the drive gears. In the position shown in (b), the latch has failed to latch because of insufficient spring tension. The portion of the mechanism in (c) is another view of the situation shown in (b).

Mechanical Problems in the Sanyo VHR9300　8

Figure 8-7. Once the faulty spring is replaced and the VCR is cycled, the brake will be in the position shown here.

Summary

Sometimes a known good unit can be the most valuable source of service information. In some cases, diagrams can be drawn during disassembly and reassembly for future reference. In most cases, the unit can be repeatedly cycled through its various modes in order to observe how the various parts interact. In time you will become familiar with many designs, each having their own peculiarities.

Chapter 9
Readout and Tape Loading Problems in RCA/Hitachi/ Sears VCRs
By Victor Meeldijk

The VCRs that were manufactured by Hitachi about eight years ago often have failed clock/readout displays. The display itself is usually not the problem, although with age it does get dimmer. This problem is usually caused by the failure of the dc/dc converter, which provides 24V for the display.

This converter is located on a vertically-mounted circuit board, called the UHF-VHF (or U-V) tuning circuit board, located behind the display circuit board (*Figure 9-1*). To remove this circuit board you may have to push a plastic latch out of the way (*Figure 9-2*) and disconnect some cables from a "circuit trap" (*Figure 9-3*).

Figure 9-1. The UHF-VHF (or U-V) tuning circuit board which contains the dc/dc converter that provides clock/display power is located behind the clock/display circuit board.

This is not a two-piece connector but a connector that clamps the cable wires. The wires are released by pushing the outside of the connector toward the circuit board. While holding the connector in, pull out the cable. The cable can be reinserted without releasing the cable trap.

9 Readout and Tape Loading Problems in RCA/Hitachi/Sears VCRs

Figure 9-2. You may have to push a plastic latch out of the way before the UHF-VHF (or U-V) tuning circuit board can be removed.

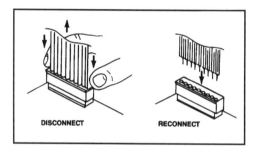

Figure 9-3. Disconnecting cables from a "circuit trap".

Models With Discrete Components

In older models, which use discrete components (*Figures 9-4 and 9-5*), a failed digital transistor Q803 (type DTC124F RCA part number 157959), electrolytic capacitors C805 (100µF, 25V; open circuited) and C807 (10µF, 16V; changed value) are commonly found. There is also a semiconductor fuse, ZD801, (marked N5S, RCA 147464, also NTE 615P) that may have opened.

Other parts that may fail are electrolytic capacitors C803 (47µF, 50V), C806 (100µF, 35V), and switching transistor Q804 (type 2SD1266 RCA part number 164231). Other parts that should be checked are the diodes, fuses, IC and power transformer in the VCR power supply section.

The semiconductor fuse looks like a TO-92 transistor with two leads, but sometimes the third lead (of the lead frame which is not connected internally) can be seen shorted to one of the other leads. A replacement device measures as a short circuit when checked with an ohmmeter.

9 Readout and Tape Loading in RCA/Hitachi/Sears VCRs

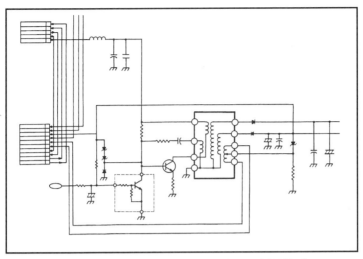

Figure 9-4. A schematic diagram of an older model VCR that uses discrete components in the dc/dc converter.

Figure 9-5. The discrete component dc/dc converter.

There may also be an open winding in the switching coil L802, (*Figure 9-6*) (RCA 164232). While this part is still available as a replacement item, you may be able to repair the defective coil. If you remove the metal cover of the transformer, using a thin flat blade screwdriver to pry around the base of the part (*Figure 9-7*), you may find that the break in the winding is right at a terminal pin. If that is the case, using a fine piece of wire, such as a single strand from a 22 gauge wire, you can bridge the gap between the winding and the pin.

9 Readout and Tape Loading Problems in RCA/Hitachi/Sears VCRs

Figure 9-6. An open winding in the switching coil, L802.

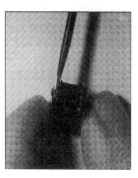

Figure 9-7. Use a thin blade flat screwdriver to pry around the base of the coil to free it from the metal shield.

Using a magnifying glass and a pair of pointed tweezers, make a small "J" shaped hook in the exposed transformer winding, do not pull on the wire while doing this as you can easily break it. Then make a similar "J" hook in the bridging wire. Hook the wires together and secure them with solder. The other end of the bridge wire can then be wrapped around the terminal pin and soldered in place. This procedure should take about 20 minutes. *Figure 9-8* shows the repaired coil.

In later designs, the discrete component dc/dc converter has been replaced by a module, as in *Figure 9-9*. Many replacement part vendors sell a dc/dc converter repair kit (RCA 163818, Hitachi 5262063) that contains an improved version of this module. The kit contains a module that has heat sink fins on it, and some replacement capacitors and a semiconductor fuse. You may find, however, that you may need capacitors in addition to those in the kit.

Figure 9-8. The coil shown in Figure 9-6 after the repair.

Figure 9-9. In later designs, the discrete component dc/dc converter has been replaced by a module. Replacement modules have a heat sink on top to improve heat dissipation and reduce failures of this part.

A Tape-stage Mechanism Problem

Often a customer will live with various VCR malfunctions, finally bringing in the VCR for repair only when a major failure occurs, such as the display problem described above. One machine I encountered exhibited several problems. When this VCR was powered up without a cassette, the cassette loaded light came on. Then the stage mechanism would move out, as if to eject a tape (the stage motor is energized) and the cassette light would go off. The stage would then try to pull in, but be stopped by the cassette loading latches (the motor would be energized and then off). During this last maneuver, the cassette light would come on again and the VCR would finally automatically power off.

9   Readout and Tape Loading Problems in RCA/Hitachi/Sears VCRs

It was possible to load a tape only when the stage tried to pull in. Attempting to load a tape at any other time caused the gears to lock up, and go out of alignment, especially if the cassette was forced in. During the time the VCR was trying to pull in the stage mechanism, if the play button was pressed the machine would go into play mode (remember this is with the stage in an up position, without a cassette).

If you just said to yourself that this should not happen because the IR tape loading sensors should prevent this, you are correct. Just by reviewing the failure symptoms, you saved a lot of time and pinpointed the failure cause. It should be noted that the cassette loading mechanism troubleshooting chart in the RCA manual does not explicitly list the sensor as a problem area. *Figure 9-10* is the troubleshooting chart from the RCA VLT 600-603 manual.

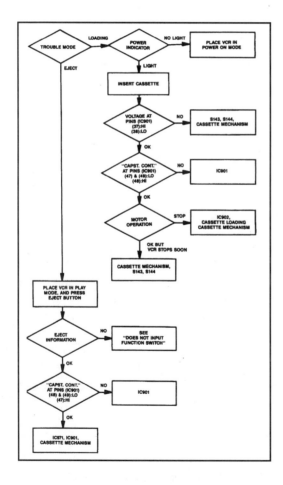

Figure 9-10. The cassette loading mechanism troubleshooting chart from the RCA VLT 600-603 manual.

9

Readout and Tape Loading in RCA/Hitachi/Sears VCRs

Solving the Tape-stage Problem

This erratic stage behavior is caused by a failed (open circuited) supply sensor which sent a message to the VCR microprocessor that there was a tape loaded in the machine. This sensor is a phototransistor with an emitter on the left side and a collector pin on the right. The base is biased by light. This particular stage design does not use a cassette-in sensor (which is located just above the tape loading area).

In this design, the presence of a cassette is sensed by the partial rotation of the tape-loading mechanism caused by the user pushing in the cassette. This partial rotation of the mechanism results in the actuation of two switches (S143, RCA 147281), the cassette-down and (S144, RCA 147281), the cassette-up.

These switches, located under the cassette loading motor, tell the microprocessor to activate the stage motor. However, with a failed end-of-tape sensor, this does not occur. *Figure 9-11* shows the switches when the right side of the stage mechanism has been removed and turned upside down.

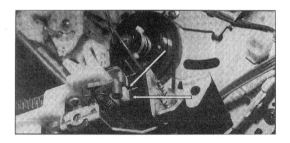

Figure 9-11. The cassette sensing switches (S143 and S144 in an RCA model VLT 600, 601, 602) when the right side of the stage mechanism has been removed and turned upside down.

There are two levers, or arms, that ride on the gears and actuate the cassette sensor switches. The left arm is RCA 162232 for VLT 600 and 161701 for VLT601 to 603. The right arm is RCA 161702. *Figure 9-12* shows the cassette-loading mechanism diagram from the RCA VLT600-603 manual.

Without a cassette, the left, or innermost arm, is down and the right arm is up. You can also see a cutout on the black gear that the arms are resting on. When a cassette is loaded, both arms are up, and the inner arm, which would otherwise be down, is held up by part of the stage-loading mechanism (the metal arm on gear RCA 162963) pressing on it.

9 Readout and Tape Loading Problems in RCA/Hitachi/Sears VCRs

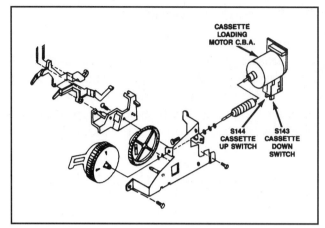

Figure 9-12. *The cassette loading mechanism diagram from the RCA VLT600-603 manual.*

Aligning the Gears

To gain access to the cassette-loading mechanism, remove the VCR top cover which is held in place by screws at the back of the VCR. Remove the front cover by taking out the front screws on the bottom of the VCR and then removing the three screws on the top of the unit. The stage is secured by two black screws in the front (where the tape is inserted) and two screws that hold it down to the VCR tape transport chassis.

When all the screws are removed, disconnect the connector going to the stage loading motor printed circuit board. The stage can then be lifted out from the rear, to clear two front tabs that are in slots in the front of the VCR chassis. The left side mechanism can be separated from the stage by removing two screws, one at the front and one in the middle of the assembly.

To align the gears, look for arrows, which have to point to each other (*Figures 13* and *14*). Be careful when aligning the gears as some of them are spring loaded and are under tension. The photographs should help you in the alignment process.

Readout and Tape Loading in RCA/Hitachi/Sears VCRs   9

Figure 9-13 and 9-14. The right side loading gears showing that they are aligned by having arrows on the gears point to each other.

Cosmetic Fixes

When all electrical repairs, and checkouts, are completed you should see if there are any little cosmetic repairs that can be easily performed (you probably remember advice like this from the business columns in this magazine). For example, in the RCA machine discussed above, the customer had the control door held in place by a piece of duct tape (*Figure 9-15*), which looked ugly. The hinge holding the control door in place had been broken off (*Figure 9-16*).

Figure 9-15. In this RCA VCR, the owner used a piece of duct tape to hold the control door in place.

9 Readout and Tape Loading Problems in RCA/Hitachi/Sears VCRs

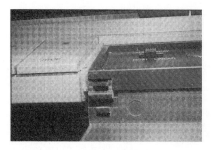

Figure 9-16. The hinge holding the control door of Figure 9-15 had been broken off.

I made a new door hinge out of a piece of thin gauge scrap aluminum, fastened and screwed it into place (I first drilled a small starter hole into the plastic and secured the screw in the hole with some epoxy, see *Figure 9-17*). This took about 15 minutes. I then cleaned the unit up and masked some scratches in the cover with some silver model paint. While these repairs did not do anything to improve the performance of the VCR they improved the appearance of the unit tremendously and the customer was pleased.

Figure 9-17. The hinge was repaired using a piece of thin gauge aluminum.

Note: Thomson Consumer Electronics can generally supply manuals for machines less than 10 years old: TCE Publications, 1003 Bunsen Way, Louisville KY 40299, 502-491-8110. For models that are almost 10 years old, or older, such as the VLT600HF (a 1985 model), photocopies of manuals can be ordered from Alexander Graphics, P.O. Box 98, 3658 Shady Lane, Plainfield, IN 46168-0098, 317-839-2372.

Chapter 10
Recommended System and Servo Control Circuit Diagnostic Procedures For VCRs
By the ES&T Staff

Adapted from a Diagnostics and Troubleshooting Techniques paper published by General Electric.

System and servo control failures account for more than half of all service performed on the electronic circuits of VCRs. Unfortunately, a great deal of time is wasted in determining the defective stage and components. Add to this the pressure of the customer or the manufacturer demanding quick service, and it becomes difficult for the technician to take the time to think the diagnosis through logically.

The purpose of this article is to reinforce some basic theory, and to suggest some logical steps in troubleshooting and diagnostic procedures. The way in which these circuits process information may differ from model to model. The results, however, are the same.

When thinking of where it makes sense to begin a diagnosis, a good approach is to think in terms of inputs and outputs.

The inputs to a device, whether a transistor or microprocessor generally are designed to make something happen, usually to an output. Assume, for example, that you suspect that the problem is a transistor switch whose function is to turn something on and off.

You push the play button and nothing happens. According to the schematic diagram, the loading motor requires 12Vdc. A multimeter measurement confirms that the 12Vdc source is supplying the proper voltage. When you measure the collector of the motor drive transistor you determine that the 12Vdc is present there. Still, pushing the play button causes no changes in the circuit.

10 Recommended System and Servo Control Circuit Diagnostic Procedures for VCRs

A Recommended Diagnostic Procedure

To track down the cause of this problem, monitor the base of the transistor with a meter. When the button is pushed, look for a change to occur. For a transistor to turn on it must be forward biased. In the case of an NPN transistor, the base must be positive relative to the emitter. For a PNP transistor, the emitter must be more positive than the base. A difference of approximately 0.6V between the emitter and base is normal for a silicon transistor and 0.3V for a germanium transistor.

If a change does occur at the base, but no change occurs at the collector, the transistor is suspect. If no change occurs at the base, check the preceding stage.

If the transistor checks good, you may want to verify the following stage and or stages by changing the input to the transistor (base), by forcing a change of state (base to ground through a low value resistor, or applying an external voltage to the base and monitoring the output: collector or emitter).

This procedure is similar to signal injection. In this case, by forcing a change of state, you cause the device to respond. This allows you to focus only on that device, and should lead to more clearly defined thinking. Failure to think logically results in incorrect diagnosis. Microprocessors have been destroyed by unnecessary desoldering. Always check the discrete components before concluding that the microprocessor is faulty. Failure to perform these checks can be costly and time consuming, and can result in PC board and component damage.

Some Diagnostic Hints

The following are diagnostic hints in troubleshooting two circuits found in all VCRs: the system control circuit and the servo circuit. Circuitry will differ from model to model. However, using a logical troubleshooting process to eliminate circuit components from suspicion will apply to all units.

Assume that you have already made the necessary checks to determine that the power supply and mechanical functions are all operating normally.

Recommended System and Servo Control Circuit Diagnostic Procedures for VCRs **10**

System Control Circuit Diagnosis

The symptom in one VCR is that nothing happens when the play button is pushed, and the cylinder motor starts then stops (in some models).

A good approach to diagnosis in this case is to check the block diagram to determine the proper signal path (see *Figure 10-1*). Locate pins 34 and 36 on the system control IC (IC6001). These are the loading motor forward, loading motor reverse control lines.

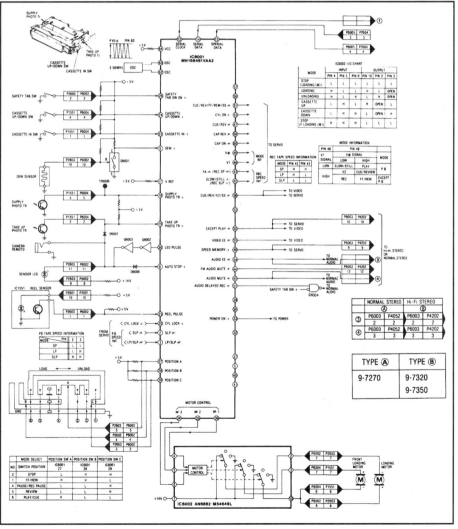

Figure 10-1.

10 *Recommended System and Servo Control Circuit Diagnostic Procedures for VCRs*

Check the flow paths on the output ports. In this case, the outputs from pins 34 and 36 of IC6001 arrive on pins 4 and 6 of IC6002. These are input ports that turn electronic switches within the IC on and off. First confirm the V_{CC} voltage on pin 7 of IC6002. Next, look for a change to occur on pins 2 and 10 of IC6002 when the play button is pushed. You determine that for the motor to load the tape, pin 10 must be high and pin 2 low. The reverse is true for unloading. If no change occurs, you recheck pins 4 and 6 while pushing the play button, remembering that the output of IC6002 will control the loading motor. To load, pin 6 will be high, and pin 4 low. If you confirm the proper change on these pins but none on pins 2 and 10, then the probability is that IC6002 is defective. In most cases, operation of the unit can be accomplished by applying 12Vdc from an external power supply to pin 8 of Plug P6009. The tape should load and the VCR should enter the play mode. Remember to remove the 12Vdc, or the loading motor will continue to run.

Now, let's assume there was no change at pins 4 and 6 of IC6002. Using an external dc supply, apply 5Vdc to pin 6 of IC6002. If the tape loads, you've determined that IC6002 is working properly. Work back on this line toward the microprocessor, checking the associated components for defects. If none appear, you might suspect the microprocessor.

Before condemning the microprocessor, you should make a few more checks. Are the inputs to other pins of IC6001 correct? Is the mode select switch functioning properly? Ask yourself these questions before you condemn a perfectly good microprocessor.

Servo Control Circuits

Servo circuitry could be thought of as the secondary operation that occurs within a VCR after the system control performs its function. Once the tape is loaded, the record and playback functions are taken over by the servo circuits. The recording of both the video and audio tracks in the record mode, and, conversely, the tracking of this information in the playback mode, must be precisely controlled for playback to be normal.

This is accomplished by two components: the cylinder motor and the capstan motor. Servo circuits can be troublesome to service since most of the circuitry contains feedback loops (information is fed back to other related circuits where it is stripped and, in some cases, reformatted and sent out again to be compared).

Recommended System and Servo Control Circuit Diagnostic Procedures for VCRs **10**

Before troubleshooting the following hypothetical problem, here is a good rule of thumb:

• If the picture produced by the VCR tends to break up horizontally, a likely cause is incorrect cylinder motor speed.

• If the video pulsates in and out at a more or less steady rate, a likely cause is erratic capstan motor speed. Most VCRs are designed to mute the audio when the servos are not locked.

There are many exceptions to these rules, but they may save you precious time and initially steer you to the correct circuitry.

Diagnosing a Servo Control Circuit Problem

The customer complained about a noise band that periodically ran up the picture when he played back a tape recorded on that machine. All other tapes, recorded on other VCRs, played back fine. When playing a tape recorded on this machine on another machine, it exhibits the same symptom.

The most likely cause of this type of symptom is a servo problem in the record mode. A quick way to confirm this would be to put a mark on the upper surface of the video head using a grease pencil. Place the machine under a fluorescent light and run it in the record mode.

Examine the spinning video head assembly. You should see a blurry, barely moving pattern of the mark you made. If you do not see this pattern, the servo is not locked. Assume that in this case you do not see this pattern, only a blur.

Let's say you also notice the following additional symptom: the video head switching point appears to move vertically through the picture. The switching point from head A to head B is turned on and off by an electronic switch. This point is chosen to be 5 to 8 lines ahead of vertical sync. It is precisely controlled by a 30Hz square wave produced in the servo system.

Take a look at the block diagram of *Figure 10-2*. Where would you suspect that the problem exists? A little reflection suggests that the problem is somewhere between points A and B.

10 Recommended System and Servo Control Circuit Diagnostic Procedures for VCRs

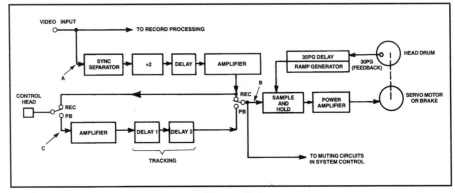

Figure 10-2.

The 60Hz vertical sync pulses from the composite video signal are separated and divided by 2. They are then amplified and sent to the control head to be used for the servo circuit. If you attempt to observe these pulses at the output of the amplifier, you will see that they are not there.

If the problem was that the servo locks in record but not in playback, this would indicate that the timing reference was lost. This would point to a failure between points C and B.

If the VCR fails to lock in both playback and record, the problem may lie with the 30PG feedback pulse generated by the video cylinder motor. Use a scope to confirm that the 30PG pulses are developed, that they have sufficient amplitude, and that they are triggering the ramp that drives the sample and hold gate. Usually, most of this circuitry is inside one IC. You can, however, trace both reference and feedback signals to points as close to the sample-and-hold gate as possible.

Chapter 11
Servicing the Hitachi VM Series Camcorder
By Timothy W. Durhan

VHS camcorders from all manufacturers have a lot in common. They have to have a lot of similarities in order to record and play back on the same VHS tape cassette. On the other hand, manufacturers also have a great deal of freedom in the details of how they design and construct their camcorders.

This article will describe procedures for servicing Hitachi models VM 3000 through VM 5000 camcorders. Many of the problem symptoms and actions to correct the problems will also apply to other brands and models of camcorder.

Hitachi manufactured thousands of camcorders in the late 80's as models VM 3000 to VM 5000. Radio Shack, RCA and Sears sold these units too, using their own names and model numbers. All feature the same tape mechanism. The capstan, mode cam and tape wind functions are actuated by belts.

Symptoms of Worn Rubber Parts

You probably know rubber parts deteriorate in time, even if they're not used often. Chances are, a five or six year old camcorder will need new rubber. Some common symptoms of worn belt problems include:

- Tape starts to load, then camcorder shuts off.
- Tape runs, then after a while, shuts off.
- Camcorder eats tapes.
- Tapes won't play or record.

If you have serviced VCRs with similar problems, you know it's not too difficult to replace worn rubber parts.

11 *Servicing the Hitachi VM Series Camcorder*

Many VCR technicians are reluctant to service camcorders, even though they wouldn't think twice about opening up and repairing a hand held remote control. If you can repair a remote control unit without destroying the case or losing any of the buttons, performing a mechanical repair on a camcorder shouldn't be too difficult. Lost screws, pinched wires and broken pc boards can be avoided by using a systematic disassembly and reassembly procedure.

Getting Started

Start by powering up the camcorder using the customer's ac adapter, since a defective battery may also be the cause of any of the symptoms mentioned earlier. Moreover, there is nothing more frustrating than running out of power in the middle of a repair. If your customer didn't include the adapter along with their camcorder, put this repair on hold until they do.

Slide the power switch to on, and press eject. If the mode belt is in good shape, the cassette lid should pop up. If it doesn't open, you'll have to trigger the carriage latch manually.

To open this latch manually, unplug the ac adapter and remove the two screws that hold the cassette lid on. Carefully slide off the lid, and set it out of your way. On the right side at the top edge of the chassis is a tiny latch (*Figure 11-1*). Gently move the latch to one side with a small screwdriver or pick. The housing should pop up, and you can remove the video tape, if one is stuck inside.

Performing the Diagnosis

Once the cassette lid is off, power-up the camcorder again. Cover up the sense LED in the center of the transport with black tape or other suitable light shield, and press play. Again, if the mode belt is in good shape, the guide posts should move to their stoppers, and the drum will start to spin.

To determine whether the tape-wind belt is doing its job, use a torque gauge on the take-up spindle. Hitachi recommends 80gm-cm to 110gm-cm. If you lack such a handy tool, you can try to stop the spindle with your fingers. Obviously, if the spindle doesn't turn, or stops very easily, the tape-wind belt is defective. Is the pinchroller turning?

Servicing the Hitachi VM Series Camcorder **11**

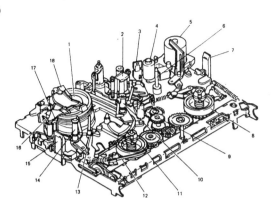

1. Upper Cylinder (Video Head)
2. Audio/Control (A/C Head)
3. Dew Sensor
4. Pressure Roller
5. Capstan Roller
6. Capstan Flywheel
7. Take-up End Sensor
8. Take-up Reel Disk
9. End Lump
10. Take-up Guide Roller
11. Supply Reed Disk
12. Tension Band
13. Tension Arm
14. Supply Guide Roller
15. Supply End Sensor
16. Impedance Roller
17. Full Erase Head
18. Cylinder Brush

Figure 11-1. Tape transport mechanism—Top view.

If the take-up spindle and pinchroller aren't moving, the capstan belt is defective, or there may be an electronic fault. Listen closely for the sound of the capstan motor spinning. You will have to press the play button continually in this condition, because the lack of pulses from the capstan and take-up sensors will alert the system control microprocessor to enter the protection mode, and the camcorder will shut off.

If you have checked all of these functions, and have determined that a belt may need to be replaced; replace them all.

There are only three belts on the mechanism, and since they were all manufactured at the same time, if one is worn, the other two can't be far behind.

Belt part numbers 6356445, 6356472 and 6358012 should be available from any Hitachi part distributor. Use numbers 174757, 174758 and 174759 if you want to order your parts from RCA instead of ordering from Hitachi.

Getting to the Belts

To begin, remove the covering from the sense LED you put on earlier. Close the cassette holder, (if you can). Unplug the adapter cable to give yourself more room. Unplug and remove the viewfinder.

11 Servicing the Hitachi VM Series Camcorder

Lay the camcorder on its side, with the cassette housing facing down, and the lens assembly pointing to your left. Remove the screws that hold the case shells together. Next, carefully pull the shell that's facing you off and set it aside.

Release the main pc board from the two white clips on both sides (*Figure 11-2*) and slide the control pc board (buttons and all) slowly toward you. Unplug the small connector from the bottom of the main pc board, and remove the large wiring harness from its holder. This should allow you to tilt the board for free access to the tape mechanism.

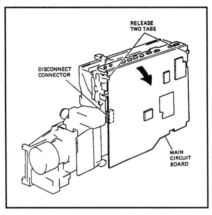

Figure 11-2. Jack circuit board removal.

Next, remove the plastic sheet covering the capstan pulley. Locate and remove the screws and cover holding the flywheel in place, and lift off the cover. Remove the old belts and clean the gum deposits off the pulleys with a solvent, such as alcohol or acetone.

Replacing the Belts

Replace the capstan belt, then the tape-wind belt. Rotate the flywheel by hand to insure that there are no twists in the belts, and remove any grease that may have found its way onto the new belts.

Reinstall the flywheel cover and screws. Reinstall the plastic sheet and inspect the area for wiring that may interfere with any movement of the mechanism. It's a tight squeeze, but you can take off the mode belt from the motor pulley and worm gear pulley without removing either one.

Servicing the Hitachi VM Series Camcorder 11

Located in the top left corner (*Figure 11-3*), these pulleys should be cleaned too. Again, make sure there are no twists or excess grease on the new belt.

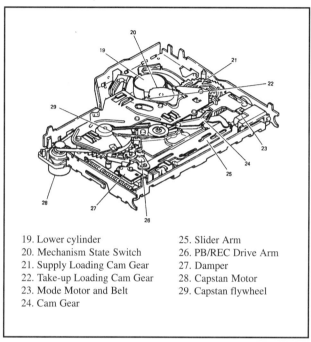

19. Lower cylinder
20. Mechanism State Switch
21. Supply Loading Cam Gear
22. Take-up Loading Cam Gear
23. Mode Motor and Belt
24. Cam Gear
25. Slider Arm
26. PB/REC Drive Arm
27. Damper
28. Capstan Motor
29. Capstan flywheel

Figure 11-3. Tape transport mechanism—Bottom view.

While you have the case off, it's a good idea to use a small soft brush to clean out the dirt and dust that has found its way inside. Reinstall the connector to the main pc board, tuck the large wiring harness back into its holder and slide the control pc board into the slots on the top case. Then snap the main pc board back into the clips that hold it in place.

The Finishing Touches

Next, turn the camcorder over and remove the other side shell. Clean the video heads, lower cylinder lip, guides, guide rollers, pinchroller, ACE heads, impedance roller and capstan shaft with isopropyl alcohol (or other suitable chemical).

Always be extremely careful when cleaning the video heads. Follow the manufacturer's directions carefully, and use only specially made plastic foam or chamois leather swabs.

11 *Servicing the Hitachi VM Series Camcorder*

Remove any excess grease and dirt with Q-tips or a soft brush. Reinstall the side shells and viewfinder.

Before powering up the camcorder, clean the lens and viewfinder window with a lens cleaning solution and lens tissue (available at any retail camera store). Then connect the camcorder and TV (or monitor) to the ac adapter and plug it in.

Slide the power switch to on, press eject and put in a tape to test play quality. If everything is in order, you should have a clear picture on the screen and in the viewfinder. Make sure that the audio is playing back at the proper level and that it is not distorted.

Perform a Thorough Operational Test

Stop and eject your play test tape and insert a tape you can record on. Remove the lens cover and put the camcorder into the record mode. While recording, use the zoom, focus and other features on the camcorder to verify that everything is working properly, and that no connectors are loose. Playback your recording and check the video and sound for accurate and natural qualities.

Because you didn't disturb any electrical circuitry or tape path geometry, you won't need expensive jigs, charts or other test equipment for a repair job such as this. Replacement of pc boards, power supply components or the CCD and associated parts would require a more involved repair and adjustment procedure.

If, after a few belt and cleaning jobs, you like the challenge that a camcorder provides, there are books available from Ryder Press, Howard Sams and others that deal with camcorder theory and operation in full detail. Also, Philips of North America, and others, have classroom education on camcorder repair.

Camcorders manufactured in the 90's require more elaborate test jigs and contain exotic concepts never thought of in the 80's. But isn't the same true in other areas of electronics?

Chapter 12
Servicing VCR Motor Problems
By Homer L. Davidson

A defective VCR motor can cause a variety of problems. It may cause the VCR to shut down, load improperly, change speeds, or fail to turn on altogether. A leaky or open motor driver IC can cause intermittent, erratic, or otherwise improper speed control, resulting in unwanted changes of tape speed in the VCR. Malfunctions in components related to the motor circuits can result in many different motor problems, which may make the service technician think that the problem is a defective motor.

A defective VCR tape loading motor may stop in the middle of tape loading, cause slow operation, or it may erratically eject the tape. When the tape will not play, suspect a loading motor. Check the loading motor when the supply reel turns for a few seconds and then shuts off. Suspect a defective motor when the tape will not unload in the VCR. A broken drive belt will prevent the tape from loading into the VCR (*Figure 12-1*).

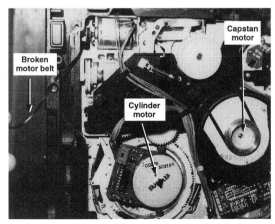

Figure 12-1. If the VCR won't load a tape, check for a broken motor belt.

Replace the defective capstan motor when you observe a change of speed in the play mode. Suspect an open motor winding, improper supply voltage, motor fuse, or a defective motor driver IC when the capstan motor will not rotate.

12 Servicing VCR Motor Problems

Usually, an unwanted change in capstan speed results from a defective servo or driver IC and speed control circuits.

It may be necessary to replace the lower drum or cylinder when the drum motor will not rotate. A VCR in which there is a defective drum or cylinder motor may load the tape and then shut down without playing. If a VCR that you are servicing has symptoms such as no drum or cylinder rotation or excessive or erratic speeds check for poor connections to the cylinder, improper voltage, or a defective main microprocessor.

Motor Supply Sources

The capstan motor moves the tape across the tape heads at constant speed. A defective capstan servo circuit or motor can affect both picture and sound in the VCR. If the sound is too fast or slow, suspect problems in the capstan circuits. Usually, the capstan speed is controlled with a speed control, phase control, and speed select circuits. Wow and flutter sounds may result from a defective capstan motor or servo circuit.

Check the voltage applied to the motor terminals. Incorrect voltage may be caused by a defective driver integrated circuit or servo section. Compare the voltage that you measure with the voltage specified on the schematic.

To determine for sure whether the problem is in the motor or in the motor supply circuits, connect an external dc voltage source to the motor terminals, to see if the motor rotates. Before connecting the external supply to the motor, disconnect at least one of the motor leads from the circuit. If the motor fails to turn with the external supply connected, measure continuity across the motor terminals to see if a motor winding or one of the terminal leads is open.

A cylinder or drum motor rotates the video heads at constant speed with respect to the tape movement (*Figure 12-2*). If the drum motor does not rotate, check the voltage across the motor terminals. Suspect a defect in the motor drive or servo IC when the drum will not rotate. Inject external voltage to the cylinder motor terminals to see if the motor begins to turn. If the motor rotates, suspect the drum motor circuits. If not, check the motor continuity with the ohmmeter.

Servicing VCR Motor Problems **12**

Figure 12-2. The cylinder or drum motor rotates the video heads at constant speed with respect to the tape movement.

The loading motor loads and unloads the tape. When the tape will not load, suspect a defective loading motor, broken belt or gear assembly, and loading motor circuits. The loading motor mechanism may be belt or gear driven. Often, the loading motor is driven by a loading motor driver IC. Check the voltage applied to the motor terminals. Apply external voltage to the motor terminals (after disconnecting at least one lead) to determine if the motor rotates or if the motor drive circuits are defective.

Loading Motor Problems

If the tape is slow in loading, ejects tape slowly, stops in the middle of loading, or if the loading motor drive IC is extremely hot, suspect that the loading motor is defective. Of course, a leaky driver IC may run red hot and cause the motor to fail to rotate. Check for a defective loading motor if the VCR will not play or load the tape. Suspect a defective loading motor or driver IC when the supply reel turns for a few seconds and then shuts off.

Check for a defective loading motor when the VCR shuts down and will not accept tape. A defective loading motor may change speeds and produce erratic loading. Intermittent load and unload may be caused by a defective loading motor or drive belt. A foreign substance jammed in the VCR opening may prevent the unit from loading.

12 Servicing VCR Motor Problems

Loading Motor Related Symptoms

A worn or cracked loading belt may be the cause of problems in loading of the tape. Failure of the VCR to load the tape may be caused by a broken or worn belt or a foreign object lodged in the belt path. Check for a defective drive belt when the tape is stuck and will not move. The VCR may operate in fast forward with no reverse action, and then shutdown if the loading belt is defective. If you hear a grinding noise when you first turn the VCR on, suspect a loading motor or drive belt (*Figure 12-3*).

Figure 12-3. If the VCR makes mechanical noise, check the drive belt, gears and loading motor.

Intermittent or erratic loading may be caused by a broken or jammed gear assembly or cassette housing. When the VCR loads and unloads intermittently, suspect a bad timing gear or loading bracket. Replace the loading motor mode cam assembly in cases where the VCR fails to rewind, or attempts to load and then shuts off.

In cases where the tape loads partially, but does not load completely up to the heads, replace the master cam gear. If the unit will not accept the tape, or ejects the tape after the cassette is inserted, replace the link gear from the loading motor to the cassette carriage.

Check the timing gears when the VCR will not load the tape. If the tape will not load correctly, check and replace the carriage link gear assembly.

If the loading motor rotates continuously, check for a leaky or shorted loading motor drive IC. If the carriage will not load and the VCR occasionally shuts off during loading, replace the motor control IC. Suspect a defective loading motor drive IC when the tape appears stuck and will not move.

Servicing VCR Motor Problems **12**

If the VCR will not load, and there is no drive voltage to the loading motor, check the motor drive IC. If the VCR will not load, rejects tape, and sometimes plays, and then goes into the rewind mode, replace the control system or servo IC. Check for a defective loading motor IC if the loading motor runs in the reverse mode but not in the forward mode.

Suspect a defective cassette control switch or defective motor when the VCR will not load or play. Check for an open fusible resistor in the motor circuits if the unit will not load. When the tape will not load, check for a defective cassette-in switch or bent cassette loading bracket.

RCA VLP900—No Loading Action

The cassette would not load in the RCA VLP900 VCR. In this unit the servo control IC controls both cylinder and capstan motors. The IC in this case was normal, so the most likely location of the trouble was in the loading motor drive circuits. Results of continuity tests at the motor terminals and at CN903 were normal (*Figure 12-4*). I measured the 9V and 12V sources at pins 1, 2, and 8, respectively. The voltages at pins 3 and 7 in the load and unload modes were not according to the specifications. I replaced motor driver IC901 with the manufacturer's exact part number (M45453), which restored the VCR to proper operation.

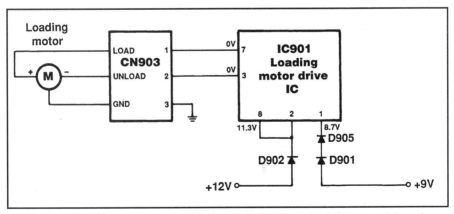

Figure 12-4. If the problem in a VCR is failure to load a tape, check the motor drive voltage. In this RCA VLP900 VCR, the motor drive voltage (measured at pins 3 and 8 on the motor driver IC (IC901) was out of specification. Replacement of IC901 with an exact manufacturer's replacement part corrected the problem.

87

12 Servicing VCR Motor Problems

Capstan Motor Problems

Failure of tape functions can be caused by a defective capstan motor or circuits. The VCR may shut down if the capstan motor is inoperative. Failure of the capstan to turn may be caused by an open motor winding, bad motor socket terminals, an open fuse, or a source voltage that is not within specification. A noisy capstan motor may require lubrication or replacement. If the capstan motor does not operate but the voltage at its terminals is within specification, check for continuity of the motor windings.

A defective capstan motor may vary in speed, or produce intermittent operation. If the capstan motor is rotating too fast, check to see if the voltage applied to the motor terminals is excessively high. If rotation of the capstan motor is erratic or intermittent, it may be necessary to replace both the capstan motor and the motor driver IC. If the driver IC is leaky or shorted, you may find that the motor is defective as well. Always replace the capstan motor with the exact part number.

Capstan Motor Related Problems

If the capstan motor does not rotate, check the motor driver IC as well as the motor itself, and replace it if it is defective. Another possible cause of failure of the capstan motor to rotate is an open fuse in the motor circuits. Measure for correct voltages at the driver and servo ICs if the capstan does not move. Replace the capstan driver IC when the capstan motor runs continuously.

If the speed of the capstan motor cannot be controlled, the problem may be a defective driver IC. Replace the motor driver IC when capstan runs for a few seconds and shuts down.

Check for open capacitors, transistors and IC in capstan motor circuits if the motor runs continually or can't be controlled. Suspect a capstan motor or the driver IC if the motor operation is intermittent, or if the motor suddenly stops rotating. Excessive capstan speed may be caused by a shorted electrolytic capacitor on the servo board. Check all components on the servo board if the capstan appears to be operating too fast.

When the VCR begins to play and suddenly stops, check for a defective mechanical position switch. You may have to replace the capstan C.B.A., when the motor will not rotate. Check for poor soldered socket connections when the motor speeds up or slows down, in either play or record modes. It may be necessary to replace a defective bracket assembly in order to correct capstan motor problems. An open capacitor in the regulated power supply can cause motor interference in the picture. Replace the defective motor with interference in the picture.

RCA VLT650HF—No Cylinder or Capstan Movement

A customer complained that their VCR didn't operate at all. Observation of operation of the unit revealed that neither the cylinder motor nor the capstan motor was turning. Because both motors were inoperative, I suspected the servo IC (IC604). I checked the +5V and +12V sources at pins 41 and 21 respectively. There was no motor drive voltage to either motor. The voltages at the loading motor IC902 pins 9 (load) and 12 (unload) were both zero. This pretty much confirmed that the servo IC was the cause of the problem. Replacement of this IC solved both motor rotation problems (*Figure 12-5*).

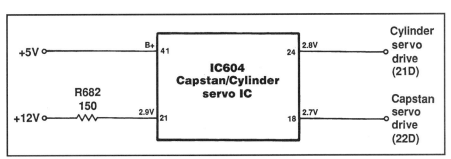

Figure 12-5. In this RCA VLT650HF VCR, the problem failure of rotation in both the cylinder and capstan motors. Voltage checks revealed that IC604 was defective. Replacement of this IC restored the VCR to proper operation.

Cylinder or Drum Motor Problems

If the tape loads and the drum rotates, and then the VCR shuts down when you place the VCR in play mode, suspect a defective drum motor. If the cylinder rotates and shuts off, check the Hall effect generator circuits and connections, before replacing the drum motor. Replace the cylinder motor if the picture jitters and you see intermittent lines at the bottom of the television screen.

In a Fisher FVH916 VCR, the VCR loads and unloads, has rewind and fast forward, but will not play. The problem is most likely an open drum cylinder P.G. coil. When the drum will not rotate, check the drum CTL voltage. If this voltage is present, troubleshoot the servo IC. You may have to replace the servo IC when the drum will not rotate.

Cylinder Motor Related Problems

If the problem is failure of the cylinder or drum to rotate, check to see if the supply voltage is correct, and check for, problems with the drum IC, or poor motor cable connections. When the cylinder motor will not start, check for poor terminal connections on driver or servo ICs. You may have to replace the driver or servo IC if there is no cylinder movement. Check for an open fuse when either the cylinder or capstan motors will not rotate. Check for open resistors in the drum circuits when the drum will not rotate. If the head will not rotate, and you measure no drive to the lower cylinder, suspect poor soldered connections in the power circuits. Check for open electrolytic capacitors in the power circuits when the drum motor will not spin.

If the cylinder motor runs too fast, or appears intermittent, check for a defective cylinder speed adjustment or a bad spot on the speed control. For intermittent operations, check for bad connections of the motor plug and socket of the lower cylinder assembly. Replace IC motor driver or servo IC when the drum motor runs too fast. Irregular cylinder speed may require the cylinder or drum circuit board assembly replacement. Check for defective diodes and transistors when the drum motor has little torque. Replace the main microprocessor when the cylinder motor rotates too fast.

RCA VLP900— No Cylinder Motor Rotation

The complaint about the RCA VLP900 VCR was that it sounded like it was operating, but there was no picture. Observation revealed that the cylinder motor wasn't rotating. Since both the cylinder and capstan motors operate from the system servo control IC801, and the capstan motor was operating, I assumed that IC801 was normal.

Voltage measurements were fairly normal on the cylinder drive motor IC505, except to the V, W, and U motor windings (*Figure 12-6*). Improper voltage was found at pins 15, 16, and 17 of IC505. I replaced IC505 with an exact manufacturer's replacement (HA13434NT). This solved the cylinder motor rotation problem.

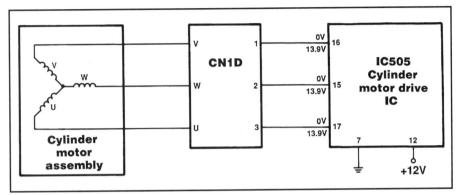

Figure 12-6. The cylinder in this RCA VLP900 VCR did not rotate. Replacement of IC505 solved the problem.

In another VLP900 VCR, with no capstan or cylinder rotation, the contacts of relay PL701 were bad. Cleaning the relay contacts solve the problem temporarily, but the correct fix for this problem is to replace the relay.

Conclusion

Although motors within the VCR can cause problems, do not overlook motor driver or system control ICs before replacing the motor. Defective reel motor circuits in some VCRs can be caused by defective zener diodes. Most motor problems can be solved by measuring voltages at the driver IC and motor, and checking continuity of the motor winding.

Chapter 13
Solving a VCR Short Circuit
By Steve Babbert

Never underestimate the value of your senses when trying to solve an electronics problem. Sometimes your senses can be of greater value than the most elaborate test bench. My senses of sight and smell recently helped me to make a repair that otherwise might have been an impossible one.

A friend brought me a Radio Shack VCR (Model 18) that had been abandoned in a house. The FCC ID number of this unit was BEJ9QK16507. These ID numbers were discussed in the December issue of ES&T (Page 8). The chart of FCC ID number prefix vs manufacturer in that particular issue shows that this VCR was made by Goldstar.

Erratic Operation

The VCR would play for a while and then shut down. Sometimes it would not play at all. I knew that it must have had a serious problem or it wouldn't have been left behind. It wasn't very old and looked good on the outside.

Because the problem was intermittent, I removed the cover of the VCR and began looking for loose connectors. Before long I found that a small board attached to the right hand side of the tape transport mechanism was sensitive (*Figure 13-1*). When I applied pressure to the board, the machine would shut down. It made no difference what mode the VCR was in. Each time this happened, I would notice a very faint odor that smelled like motor brushes sparking; however, I couldn't see or hear any spark discharge. Every time the unit shut down, I had to reset the controller IC by first unplugging the VCR and then plugging it back in.

13 Solving a VCR Short Circuit

Figure 13-1. Here you can see the circuit board mounted on the right hand side of the tape transport mechanism. Any pressure applied to the board or the connecting wires would cause the unit to shut down.

Someone Else Had Been Here Before

I had to remove the tape transport mechanism to get a closer look at the small circuit board. As I was doing this, I noticed that some plastic screw wells were stripped, probably as the result of repeated disassembly and reassembly. At this point, I was pretty sure that someone else had been on the same track I was on.

With the tape transport mechanism removed, and in its fully unloaded position, I marked the gears with a permanent marker to insure proper reassembly. Now I could remove the motor, worm gear, and idler gear that covered the board. Finally, I removed the board itself. Close inspection of the board revealed nothing unusual, and continuity checks showed no open traces or bad connector contacts.

The only component on the board was the Take-Up Reel End Sensor. Other than this sensor, the only things on the board are the traces that route signals from the tape sensors and switches to the system control. The next step was to power up the machine with the board removed from its mounting support. The purpose of this was to help localize the problem.

The VCR did seem to power up with the board removed. I repeatedly flexed the board and worked the connectors, but it never shut down when the board was removed. I even applied pressure to the mounting support and surrounding area, and again, it never shut down. Had I dislodged a particle from a connector

Solving a VCR Short Circuit **13**

when I removed the board? Maybe I broke through some oxide on one of the pins. I touched up the solder connections, to be safe, and then reassembled the entire unit.

The Problem Reappears

I put the VCR into the play mode and everything operated perfectly. I applied pressure to the board, and just as before, the machine shut down. Again there was the faint smell of a spark discharge. Now I understood why the screw wells were stripped; someone had been down this road several times. Very possibly the unit had been diagnosed as too costly to repair.

At this point, I became more determined than ever to locate the source of this strange problem. I realized I had to find this problem with the unit assembled. I moved my lights around so that I could better view the area from different angles, and finally, I got lucky.

The next time I caused it to shut down, I saw a small spark between the mounting tab at the top of the board and the metal mounting support that held the board in place. Even with my 20/20 vision, I saw the faint spark only because a shadow covered the area. Now I knew that one of the traces was coming into contact with the mounting support, but I didn't know how this was possible?

Once again, there was no choice but to remove the board from the transport assembly, but at least now I knew exactly what section of the board to focus on. I removed the board and placed it under a good light that revealed a type of design flaw that I'd never seen before. A circuit trace was routed along the tab of the board that fits into the metal support. There was only a thin layer of enamel to insulate this trace from ground. A very small burned spot is visible in *Figure 13-2*.

The purpose of this trace is to route V_{cc} to the Tape End Sensor. When this trace came into contact with the metal support, V_{cc} was pulled to ground. No doubt a fail-safe circuit caused the unit to shut down.

When the load motor was actuated, friction eventually scraped away a bit of the insulating enamel. since there was a lot of mechanical movement in this area. I wondered how often this problem had occurred in VCRs using this transport mechanism. I had a Sears VCR on another bench that used the same mechanism, but it showed no signs of this problem.

13 Solving a VCR Short Circuit

Figure 13-2. With the board removed, close observation will reveal a tiny burned spot where the trace came into contact with ground.

Figure 13-3. With the unit reassembled, and the paper insulator in place, the problem was corrected. Here are two views of the PC board with the insulating cardboard in place.

Correcting the Problem

There are any number of ways that one could correct this problem, but I chose the least costly and time consuming route. I reassembled the machine, and folded a strip of paper cut from a business card. You can see the folded strip in *Figure 13-2*. I then slipped the paper insulator between the board and the support. Next, I folded the ends over the mounting support to keep it from slipping down through the support opening. The paper insulator would be held in place by the wires once they were routed into their normal position. Now I make this modification whenever I service a VCR that has a chassis of this style.

Chapter 14
Solving VCR Servo System Problems
By Arthur Flavell

Servo systems are essential to producing stable, noise-free video from a VCR. Their job is to regulate the rotational speed of the cylinder and capstan motors, and to keep them synchronized with a stable reference signal. Servo problems may cause symptoms such as jitter in the picture, loss of sync, or noise bands in the reproduced video.

Servicing servo circuits in a VHS VCR requires an understanding of both circuit operation and VHS fundamentals. This discussion will familiarize you with the components of servo systems and their operation in the VHS format.

See the sidebar for an analysis of the components of the servo system of a VHS video cassette recorder.

VHS Format

The VHS format uses three separate areas of the video tape on which to record and play back information (*Figure 14-1*). A linear track along the upper edge of the tape contains the audio information. Video information is found in the center portion of the tape on helical tracks. Each track contains one field of video signal. A linear track along the bottom edge of the tape contains CTL, or control track, synchronizing pulses.

To ensure compatibility with NTSC (National Television System Committee) video systems and interchangeability with other VCRs, the timing of the signals recorded on tape must be precise and consistent. The electronic circuits of the VCR control the timing of signal generation, but constant tape velocity is necessary to maintain the timing in the finished videotape recording.

14 Solving VCR Servo System Problems

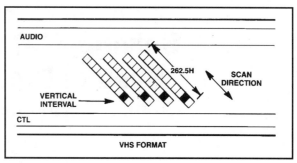

Figure 14-1. The VHS format uses three separate areas of the video tape on which to record play back information: a linear track along the upper edge of the tape contains the audio information, video information on helical tracks in the center portion on the tape and a linear track along the bottom edge of the tape contains CTL, or control track, synchronizing pulses.

Controlling tape transport speed is the job of the capstan servo. The capstan must move the tape at a constant speed so that the audio and CTL tracks record and play back properly. It must also ensure that the video tracks are aligned under the video heads for read and write operations.

The cylinder (or drum) servo is responsible for maintaining the correct rotational speed of the drum containing the video heads. This ensures that the correct number of horizontal lines are recorded or played back on each track. In addition, it must position the heads so that read and write operations begin and end at the proper place along the track.

Servo Operation During Playback

In simplest terms, the servo system is a feedback loop control circuit. It senses the condition of the operating system, compares it to a reference and generates a correction signal to maintain synchronization. Operation of the servo system is slightly different in playback and record modes. Let's look at the playback function first.

The capstan servo circuit uses two comparators to control motor speed. The speed comparator compares the V SYNC and capstan FG signals. The FG signal is amplified, processed by a Schmitt trigger, and applied to the comparator. The comparator's output is used to control a pulse width modulator.

Solving VCR Servo System Problems 14

The PWM's square wave signal is fed through an integrator, which smoothes it into a dc voltage proportional to the duty cycle of the PWM output. This dc voltage is then applied to a summing amp and routed through the CAP ON switch to the capstan motor driver. If the comparator detects a slowing of the motor, the duty cycle of the PWM is increased, providing a higher dc output from the integrator and greater drive to the motor. If motor speed is too high, the duty cycle of the PWM is decreased, resulting in reduced drive.

The second comparator in the capstan circuit is the phase comparator, which acts as a fine speed adjustment. It produces an output that controls a PWM. The PWM output is processed by an integrator and applied to a summing amp, where it is combined with the signal from the speed comparator. The capstan phase comparator uses several inputs to make the fine adjustments; V Sync, the output of the tracking multivibrator, CTL and the head switching signal.

The cylinder motor servo also uses two comparators, and operates in the same manner as the capstan circuit. The cylinder speed comparator uses V Sync and the cylinder FG signal to provide coarse speed control. Fine speed control is accomplished by operation of the CYL/ CAP phase comparator.

The H SW GEN is driven by cylinder PG pulses. The output of the generator is a 30Hz square wave which is used by several other circuits. In the playback mode, this signal supplies inputs to the CAP/ CYL phase comparator, the TR MM GEN, the head amp select circuit and the V-lock gen.

The head amp select circuit sends a switching signal to control operation of the playback head amplifiers. The head amps are alternately turned on and off so the signal from the head that is currently in contact with the tape is turned on. Each head reads alternate fields of the video signal. When head switching occurs, video noise is produced. The timing of the switch from one head to the other is critical to stable picture production. *Figure 14-2* shows the proper timing relationship.

Normal switching in the VHS format takes place 6H ±1H before the beginning of vertical sync. H is the designation for one horizontal line. This places the switching event in the overscan portion of the picture, off the bottom of the screen.

14   Solving VCR Servo System Problems

If head switching takes place too early, the switch point will be visible in the lower part of the picture. If switching takes place too late, it can interfere with reproduction of the vertical interval, resulting in vertical jitter, jumping, or loss of vertical sync altogether.

The V-lock gen provides a false vertical sync signal which is used during search mode operation. NTSC televisions or monitors can only reproduce video at a 30Hz frame rate. If the vertical signal strays too far from this figure, vertical sync is lost and the picture rolls. In search mode, vertical sync information from the video tape is not at the correct frequency and cannot be used to lock vertical scanning. The V-lock generator takes over during these periods to produce a picture that is stable.

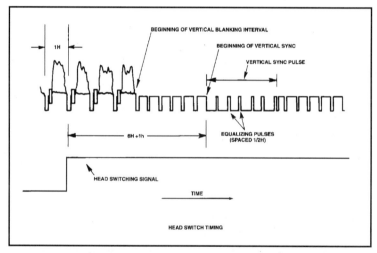

Figure 14-2. The head amp select circuit sends a switching signal to control operation of the playback head amplifiers, which alternately turn on and off so the signal from the head that is currently in contact with the tape is turned on. The proper timing relationship for head switching is shown here.

Servo Operation During Record

Record operation of the servo circuits is similar to playback operation with three major differences. The first is the source of the V-Sync reference. In record mode, vertical sync information is taken from the incoming video. A sync separator and shaper circuit in the video section produces the signal.

Solving VCR Servo System Problems **14**

The second difference is a change in one of the reference inputs for the CYL/CAP phase comparator. Because the tape is in the process of being recorded, no CTL signal exists to be used as a reference. The capstan FG signal is counted down to provide the necessary reference. The select switch operates to connect the FG signal to the comparator when the VCR is in record mode.

The third difference is in the CTL head circuit. When the VCR is in the record mode, the CTL amp is turned on and passes a signal from the CTL GEN which is recorded on the CTL track. At the same time, the PB ON switch opens.

Troubleshooting

Symptoms of servo troubles, such as jitter, sync problems, or noise bands in the picture, are often produced by mechanical components in the VCR. Before troubleshooting electronic circuits, be sure to check common trouble areas first. These include: dirty or blocked video heads, dirty CTL head, tape slippage caused by dirty capstan and pinch roller, or improper tape tension because of improper adjustment, dirty or worn drive belts or dried-out lubricants on the reel spindles.

The next step is to determine if the servos are malfunctioning in the playback mode, in the record mode, or both. Play a test tape and check for symptoms in the video to check the playback mode. Record a test signal and play the tape back in a known-good VCR to check the record mode. If the symptoms only appear in one mode, concentrate on the circuit elements that are exclusively related to that mode.

Determining if a symptom is caused by a capstan problem or a drum servo problem can sometimes prove difficult. A quick way to check for off-speed operation is to measure the frequency of the FG signal from each motor. The service manual contains specifications of proper signal frequency and amplitude.

If you find a discrepancy, perform the electrical alignments for the servo circuits. Depending on the age and model of the VCR, these may include: head switching position, tracking set, capstan free run and cylinder free run. If the V-Sync frequency is off in playback mode, alignment of the system control VCO may be necessary. If the trouble symptom persists after alignment, troubleshoot the servo circuits for component problems.

Troubleshoot in This Sequence

Begin troubleshooting by checking supply voltages for the servo control IC and the affected motor drive. Voltages should be checked for proper amplitude and absence of noise and ripple. If problems exist, troubleshoot the power supply circuits.

Check the control inputs to the servo control IC. These include: V-Sync, tracking control, PB FM signal, CTL (in playback mode), capstan FG, PG shifter and cylinder FG/PG. These should be checked for proper amplitude, frequency and waveshape as specified in the manual. If an improper or missing signal is noted, troubleshoot the signal's source circuits.

Check the control outputs from the IC. These include: head amp select, false V-Sync, CTL (in record mode), R/S/F and the outputs of the capstan and cylinder PWMs. If all inputs are correct and faults are found in one or more outputs, isolate the defective output before concluding that the IC is at fault. To perform this check desolder the appropriate pin and check the output in an unloaded state. If the output signal is now present, check the circuit fed by that pin for faults. If the output is still not present, the control IC is defective.

If all outputs from the servo control IC appear normal, check the peripheral circuits for proper operation. These include: the integrators, sum amps, capstan and cylinder on switches, motor drivers and the motors.

Breaking the Loop

Occasionally, you will find a situation where all inputs and outputs from the control IC appear to be normal and all peripheral components seem to be operating properly, yet the trouble symptom persists. This frustrating state can usually be traced to the fact that the servo system is a loop. One section of the system may be correcting in the right direction, but over-reacting. This causes the circuit to "chase its tail." To troubleshoot this type of problem, it is necessary to break the loop and observe the individual circuit elements to determine which of them is over-reacting.

A convenient spot to break the loop is the input to the motor drive circuit. Using a variable dc bench supply, apply the normal control voltage as specified in the service manual. Vary the voltage slightly above and below normal. Ob-

Solving VCR Servo System Problems **14**

serve the response of the PWM outputs, the integrator outputs and the sum amp output as the motor speed changes.

VCR servo circuits are sometimes troublesome for technicians. Because of their reliability, we seldom have to deal with them and they may not be as familiar to us as other circuits. A scarcity of technical information on the servo operation also contributes to the problem. Perhaps this information will help you solve servo troubles you may encounter and reduce frustration at the bench.

Chapter 15
VCR Mechanical Problems
By Philip Zorian

During a two hour movie a VCR pulls about eight hundred feet of videotape through its tape transport mechanism; twice that amount if you rewind it. Since the rubber belts, pinch roller, and idler tire are directly involved in the moving of this tape, their wear causes many of the problems about which VCR owners complain. Most of the mechanical problems are caused by worn-out rubber parts.

This article will first discuss replacement of rubber parts and then closely examine ten of the more common problems associated with these parts (the belts, pinch roller, or idler tire) wearing out.

Replacing Belts; Don't Measure Them

Once you've determined that belts need to be replaced because they are loose, slipping, worn, shiny, or broken what do you do next? Well, in order to replace any rubber part on any make or model, you'll need to know the part numbers, and you'll need a reliable distributor. Here are two rules I always follow when replacing worn belts on a VCR.

Rule 1: Don't measure the belts. There are two good reasons for this: first, measurement is a real time waster, and second, it is not an accurate way of determining the replacement part. In order to be measured, a belt must first be removed from the unit. Since removing the belts often requires the dismantling of the VCR, time is wasted in having to take them off in the first place, and now you've got a partially disassembled machine on your hands. It is much easier to look up the exact part numbers, call out with your quote, and then move on to the next defective machine. (I'll explain how easy this is in a minute.)

Trying to measure a worn-out, stretched, and possibly broken piece of rubber is not only frustrating; but it does not produce an accurate measurement. It's hit-or-miss, especially since most specifications for belts and idler tires are given

15 VCR Mechanical Problems

in thousandths of an inch (or millimeter). Even though some of the distributors will try to get around this by providing pictures to match your part up with, I would much rather consult a source that can tell me the exact part number needed for a belt or for the idler tire (which is even more difficult to measure accurately.)

One source of information on belts is the 1996 PRB LINE Cross Guide from Premium Parts+, Whitewater, WI.

Fortunately, pinch rollers are easy to measure accurately in a matter of minutes. First, the roller must be removed, which is fairly simple in most VCRs. Then use an inexpensive plastic caliper (for example, part #21-1220, available from MCM Electronics, Centerville, OH) to match the specifications found in the PRB catalog under "VCR Pinch Rollers".

To match specifications: (1) measure the height of the pinch roller; (2) measure its outside diameter; (3) determine the inside diameter of the bearing by measuring the width of the shaft on which the pinch roller sits. (Don't despair, it's either 3mm or 4mm, and you can eyeball them after a while.)

Since you cannot measure the inside diameter of the bearing down inside the pinch roller, you must measure what the pinch roller sits on, in other words, the shaft. Remember, the pinch roller, working together with the capstan shaft, is responsible for pulling the videotape through the machine at a precise speed. If the roller is dried out, cracked, slick, or pitted. It will cause video problems.

Rule 2: You must find a reliable distributor for these parts. The ideal distributor will carry every belt, idler tire, and pinch roller for the most popular VCRs.

The Ten Most Common Problems

The preceding paragraphs discuss replacing the rubber parts found in any VCR. Now let's take a look at some of the more common symptoms that indicate these parts have worn out. My list of the top ten symptoms are as follows:

- incomplete loading
- eating of the tape
- tape slippage

- no rewind
- no fast forward
- squealing sounds
- take-up reel won't turn
- intermittent shutoff
- tape-edge damage
- dead unit

Incomplete Loading

When the problem is incomplete tape loading, the customer will usually complain that the VCR just shuts off. Remember, all symptoms must be verified in order to properly diagnose the problem. In this situation, there is an attempt to load a tape, the unit accepts the tape and you hear the cassette drop into place, but then it shuts down.

With the "hood up" you will see the guideposts attempting to load the videotape around the video drum, but they will only get halfway before the unit shuts down. You may also hear the loading motor continue to spin.

Many VCRs have small loading belts that will begin to slip from wear and cause this symptom. Also check the guidepost tracks: if they are dry, they will need to be cleaned and lubricated.

Guidepost problems may have led to the slipping of the belts in the first place. If you don't lubricate the tracks, the new belt may not last long.

Eating the Tape

It's fairly common to find that a VCR has eaten a tape. It is important to determine exactly what is meant by "eaten," since there are a few different problems that are described in this way. If the tape is not drawn back into the cassette completely by the supply reel when the eject button is pushed, but left hanging out, it will get crunched because the cassette flap will close on the tape as it emerges from the VCR. I have found that a clear videotape is indispensable for observing this symptom. This problem is almost always caused by a worn-out idler tire.

15 VCR Mechanical Problems

Tape Slippage

Remember that the pinch roller and capstan shaft are responsible for pulling the videotape through the entire playback/record mechanism. They literally grab the tape by pinching it between roller and shaft. Over time, the pinch roller will become slick and hard as the oxide coating from the videotape adheres to its surface. As the roller ages, it loses its ability to grab the videotape. A new roller has a dull black rubber surface that is firm but supple.

When the pinch roller cannot pull the tape through at a precise speed, the following symptoms will occur: (1) the take-up reel tries to pull the tape through instead, but the speed is not precise and the audio and video will intermittently speed up. (It is usually quite noticeable in the audio; you'll get a sudden "Mickey-Mouse" sound.) (2) The opposite can also occur; the movement of the tape will slow down, and you will hear a low, deep, dragging voice. (3) If the take-up reel/idler-tire combination is unable to pull the tape through, the unit will usually shut down.

The solution to this problem is simply to replace the pinch roller.

No Rewind

The refusal of the VCR to rewind is one of VCR owners most common complaints. The most likely cause of rewind problems is a worn idler tire. During rewind the idler shifts over to the left and engages the tire against the supply reel. But if the tire is too badly worn, it is unable to create enough torque and the rewind is either weak or non-existent.

If a VCR can't rewind at all, the unit will typically shut down. This can also show up as either a weak fast forward or the loss of fast forward, since this mode causes the same idler to shift to the right in order to engage with the take-up reel.

Squealing Sounds

When a customer complains about squealing sounds, the squealing is most likely being caused by slipping belts. This is usually the first symptom before

VCR Mechanical Problems

the VCR begins to display more serious problems. If I hear squealing sounds, I will usually assume one or all of the belts need to be replaced, but the squealing sounds are typically caused by loading belts.

Since the loading motor spins suddenly and at a high rate, slippage is a common problem with these belts. You should check the belts by looking for a shiny surface where the belt meets the pulley. Also, try turning one pulley while holding onto the other connected to the same belt: if you are able to turn the pulley easily, then the belt is slipping. A new belt that is tight will make it difficult to turn one pulley if you are holding the other. Squealing sounds described by a customer can also be caused by a dry capstan shaft or motor shaft that simply needs lubrication.

Take-up Reel Won't Turn

Failure of the take-up reel to turn in a VCR will usually result in automatic shutdown. When this type of shutdown occurs, the complaint from the customer will usually be that the VCR quit working. But during observation with the hood up, you will notice the movement of the take-up reel is weak or that it is not turning at all, causing tape spillage just after the pinch roller. This is almost always due to a worn idler tire.

Once the pinch roller is finished with the tape, the take-up reel pulls the videotape safely back into the cassette. If the take-up reel is unable to turn, the sensors will cause the VCR to shut down in order to avoid damage to the tape or the VCR due to the internal tape spillage.

In newer model VCRs, the idler tire has been replaced with a gear. In these newer models, unless some of the teeth on the gear are missing, when there's a shutdown problem, you can rule out the possibility that it was caused by failure of the take-up function. However, most VCRs still use an idler tire. These tires become slick, dried, and cracked along the outer surface and must be replaced.

Intermittent Shutoff

Intermittent shutoff can be caused by many things, some of which are beyond the scope of this article. But, since it is an intermittent problem, you should consider the belts, idler tire, and pinch roller to be suspect. One approach is to

15 *VCR Mechanical Problems*

replace these parts and then bench test the unit for three hours. This should solve the problem in most cases, especially if you observe any of these parts to be worn out.

Tape-edge Damage

Tape-edge damage can be observed by opening the flap of the videocassette and noticing a curling on either the lower or upper edge of the tape. A brand new tape will appear perfectly flat and shiny.

Sending a tape with edge damage through the VCR will wreak havoc on the video, causing all kind of symptoms: picture roll, jitter, poor tracking, bending, etc. This segment will focus only on damage to the edge of the tape, but keep in mind that a videotape can be damaged anywhere along its surface.

The only rubber part that can cause tape-edge damage is the pinch roller. There are other, non-rubber, parts that can cause this problem, but an old and damaged pinch roller is my first suspicion when I see tape-edge damage.

When the roller gets so old that it is no longer perfectly flat, it applies an uneven pressure when it presses the videotape up against the perfectly flat capstan shaft. With the hood up you can easily verify exactly where the damage is occurring.

Using a tape that has no damage, run it through the unit in play mode for five seconds, and then hit pause. Using a tiny brush and white paint, put a mark on the inside non-playing surface of the tape wherever the tape contacts a part in the tape transport, then remove the videotape from the unit. The damage will always occur at some point along the path, and your mark will indicate which part is to blame. Always check to see if tape-edge damage is the cause of video problems.

Dead VCR

There are times when a belt will either break from age or simply jump off the pulley. The symptoms that result are too numerous to list and would depend on the belt and the particular machine. But a belt falling off its pulley and getting caught up in the gears will almost always result in a blown fuse as the unit struggles to overcome the obstacle by drawing more power. When the fuse blows, the customer will complain that the unit seems dead.

Conclusion

In this article I have covered the replacing of the rubber parts found on VCRs, and the ten most common problems resulting from worn-out parts. This information should equip you to deal with most problems reported by VCR owners. However, I cannot emphasize enough the importance of verifying the symptoms that customers complain about with your own eyes and ears.

Chapter 16
VCR Servo Problems: The Diagnostic Device Revisited— Part 1

By Steve Babbert

An article, "Servicing VCR servo systems," by Stephen Miller, published in the October 1988 issue of this magazine, outlined a method of troubleshooting servo circuits by voltage substitution. The article included plans for building a diagnostic device that could supply a precisely adjustable voltage to substitute for the servo's motor control voltage.

The device uses two ten-turn potentiometers for coarse and fine voltage adjustment. A three-position switch enables the user to raise or lower the voltage for kick-start and stop without having to adjust the controls. See the schematic diagram for this device in *Figure 16-1*.

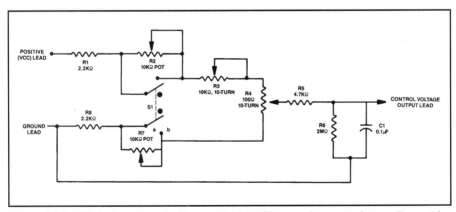

Figure 16-1. This is the schematic diagram for the VCR servo diagnostic device. R_3 provides coarse speed adjustment while R4 allows fine speed adjustment. R_2 and R_7 provide adjustments of minimum and maximum voltage. S_1 is a DPDT center-off switch that provides three operation modes: kick-start, run and stop. R_1, R_5 and R_8 provide current limiting during kickstart operation. The network consisting of R_5 and R_7 and C_1 forms a current limiting filter.

16   VCR Servo Problems: The Diagnostic Device Revisited—Part 1

This article will show how voltage substitution helped to find the cause of an unusual problem in a recent model VCR. First, let's look at why servos are needed in a VCR, and how they work.

The Helical Scan System

Because of the high frequencies associated with the video signal, conventional, or "linear," recording techniques are not practical for video. In magnetic tape recording, as the frequencies increase, the head gap must become smaller or the tape speed must become greater, or both.

Physical limitations in manufacturing place restrictions on how small a head gap can be. High tape transport speeds translate to longer tapes requiring larger reels. This problem is eliminated by using the helical scan method of recording and playback for video.

In helical scan recording, the tape traverses a helical path around the drum containing the video heads. As the tape is pulled around the rotating drum during recording, the head records on it diagonal tracks which are very closely spaced. Using this method of recording, a single inch of tape will contain the equivalent of many feet of video track information.

In the basic system, there are two video heads spaced 180 degrees apart on the drum. The azimuth of each head is shifted by a few degrees to prevent crosstalk in the event that one head overlaps an adjacent track. Narrow guard bands are inserted between tracks to help prevent this problem. For faithful reproduction during playback, a means must be provided whereby the video heads will remain on their respective tracks.

The Servo System

In essence, servos are self-correcting feedback loops in which a system can monitor its own actions and provide any needed correction. The capstan servo, which controls tape movement, consists of two loops: speed and phase.

The Speed Control Loop

The speed control loop, which can be considered to provide coarse speed correction, functions by comparing divided down FG (frequency generator) pulses to a stable reference frequency in an AFC (automatic frequency control) circuit. Though there are different systems in use, one common method uses 29.97Hz as the stable reference. It is derived through division of output of the 3.58MHz oscillator.

The FG pulse, which is induced in a stationary head coil during record by magnetic pole pieces in the capstan flywheel, has a frequency that is directly proportional to the motor's speed. In some machines, the FG head is replaced by a "Hall effect" device.

The FG pulse train is divided before being compared to the reference. This is necessary because the FG frequency is much higher than the reference frequency. Furthermore, since the FG frequency depends on the motor's speed, it also depends on the speed at which the tape was recorded (SP, LP, or EP).

One popular frequency generator design uses 240 magnetic pole pieces in the capstan flywheel. The FG frequencies are approximately 719.3Hz (SP), 359.7Hz (LP) and 239.8Hz (EP). These frequencies, which, incidentally, are all multiples of 29.97, will be divided as needed to obtain the correct comparison frequency. Nonstandard division rates may be used during special effects as directed by the system control.

Division takes place in a variable division block which is controlled by the auto speed select circuit (to be discussed shortly). Any difference between the frequency of the divided down FG pulses and the internal reference causes the speed PWM (pulse-width modulation) block to adjust its duty cycle (the ratio of on time to off time). Longer on times render higher average dc values.

The Phase Control Loop

The phase control loop, which can be considered to provide fine speed, or angular position control, functions similarly to the speed control loop. Its purpose is to retard or advance the position of the tape with respect to the video heads by causing momentary increases or decreases in the motor's speed as needed. It compares the CTL (control) signal to a stable reference in the APC (automatic phase control) circuit.

16 VCR Servo Problems: The Diagnostic Device Revisited—Part 1

The CTL pulse is read from the control track on the lower portion of the tape by a stationary head. The control track utilizes the linear recording method which is used in conventional audio recording. The tracking control provides an adjustable delay in the phase of the CTL pulse with respect to a reference.

The Tracking System

The tracking system is used to compensate for mechanical variations between different machines. The same effect could be achieved by changing the physical position of the CTL head with respect to the head drum. Any difference in phase between the CTL pulse and the reference causes the phase PWM block to adjust its duty cycle.

The speed and phase PWM outputs are low-pass filtered, or "integrated," leaving only the average dc value, which is proportional to the duty cycle. These two voltages are then matrixed and amplified before being applied to the capstan motor drive circuit to control its speed and phase.

It's worth noting that this control voltage is not the motor supply voltage. The control voltage causes the motor drive circuit to supply the appropriate voltages to the motor windings.

Head Drum Speed and Phase Control

The head drum, or "cylinder," servo also uses a speed and a phase loop. It functions much the same as the capstan servo. The main difference is that both FG and PG (phase generator) heads, in conjunction with magnetic pole pieces in the rotating head drum, are used to provide the required feedback. The drum servos are simpler because the drum speed doesn't change for different tape speeds or special effects. A good source for more information on how servos work is Sencore's Tech Tip 176.

The Auto Speed Select Circuit

The auto speed select circuit is used to determine the speed at which the tape was recorded. The CTL pulse is the key to the operation of the auto speed select circuit.

VCR Servo Problems: The Diagnostic Device Revisited—Part 1

During record, the CTL pulse is 29.97 Hz, regardless of the tape transport rate. During playback, the CTL frequency will only be correct if the tape speed is the same as it was during record. By measuring the CTL pulse, the auto speed select circuit can determine if the tape is traveling too fast or too slow. Then this circuit can cause the FG pulses to be divided accordingly.

The divided frequency, when compared to the reference, will cause the speed PWM to adjust the capstan speed. When this speed reaches a point where the divided down FG pulse frequency matches the reference, the AFC circuit will cause the PWM block to hold its present duty cycle. This will keep the motor at its correct speed.

Solving a Capstan Speed Problem

A customer brought in a Sears Model 580.53295750 VCR that was exhibiting a capstan speed problem. The owner brought it to me for a second opinion after being told by another service center that the capstan motor was defective. Their estimate was too high to justify the repair.

When I ran a quick check to observe the problem at first hand, the capstan motor was wildly erratic in the play mode but not in the fast forward or reverse modes. Based on this, I was pretty sure that the problem was more likely in the servo circuits. A defective motor generally will not run properly in any mode. I gave the owner a lower estimate, based on my evaluation.

When the unit was placed in the play mode, the video and sound would speed up and then abruptly stop two or three times per second. A clear picture was always visible on the monitor but there were noise bars in the picture that are characteristic of the search modes. Occasionally the speed would become normal for about a second.

The speed at which the tape had been recorded made no difference. The symptom was also present when using a tape test jig. Since the test jig has no tape, there was no possibility that the symptom was caused by mechanical problems in the tape path.

16   VCR Servo Problems: The Diagnostic Device Revisited—Part 1

An Attempt With No Schematic

I did not have a schematic for this model, so I looked for the pinouts of all ICs relating to servo operation in my cross references. Unfortunately, none of the ICs were listed. I was going to have to rely on my understanding of the circuit and whatever common test points I could find. Fortunately, the servo section boundaries were clearly marked on the PC board.

My first step was to check the capstan motor speed/phase control line, which originates at the servo IC and ends at the capstan motor board after passing through the integrators. At the motor board this line is yellow, whereas all other lines are gray (*Figure 16-2*). Measuring the voltage of this line while watching the analog bar graph of my DMM, showed that it was jumping erratically in step with the motor.

Knowing that this control voltage is derived from the combined outputs of both the speed and phase PWM blocks within the main servo IC, I attempted to backtrack. This unit uses just one 42 pin IC for both the capstan and drum servos. Two other 14 pin ICs, which were identical, were op amps (operational amplifiers).

I was able to trace the erratic control voltage back to pin 15 of the servo IC, which I assumed was either the capstan speed or phase PWM output. I could find no meaningful waveform at this pin with an oscilloscope.

Figure 16-2. The capstan motor speed/phase control line originates at the servo IC and ends at the capstan motor board after passing through the integrators. At the motor board this line is yellow, whereas all other lines are gray.

16 VCR Servo Problems: The Diagnostic Device Revisited—Part 1

Checking For a Missing Signal

I considered the possibility that one of the feedback signals or the 29.97Hz reference was missing or intermittent. Because the head drum appeared to be running normally, I ruled out the possibility that the 29.97Hz reference signal was missing. This signal is necessary for the proper operation of both the head drum and the capstan.

Just to be sure, I used the frequency counter function of my DMM and hunted for this signal. I found it at several pins on the op amps and one pin of the servo IC. I eliminated the possibility of a loss of CTL pulse from the control head, because the problem occurred even when using a cassette test jig. Again, because the test jig contains no tape there can be no CTL pulse.

I observed the capstan FG pulse at the FG head using the oscilloscope, but its amplitude and frequency were unstable because of the capstan speed variations. I was going to have to open one or more loops to isolate this problem. As in any loop system, a problem in one part of the loop will affect measurements in all other parts. The VCR diagnostic device would be helpful here.

Opening the Loop

When I built the diagnostic device (*Figure 16-3*), instead of using voltage from the VCR for the supply, as was done in the original article, I decided to use a 4.8V Ni-Cd cordless telephone battery. This way, I would only have to make two connections to the VCR: one to the motor's drive control line and the other to ground. With the fine adjust control set to its center position, I adjusted the coarse adjust control for an output of about 2.3V. This is a common "run" voltage.

I traced the capstan motor control line to the component side of the main PC board, in order to gain access to it while the machine was right side up. I opened the line and connected the device's positive lead to the motor side of the line and the negative lead to the VCR's ground (*Figure 16-4*). I left the line that normally supplies the control voltage from the servo circuit unconnected.

Using a tape recorded in the SP mode, I placed the VCR into the play mode. When the tape was fully loaded around the head drum, I flipped the switch into the kick-start position and then quickly back to the run position. The capstan motor was now running at a constant speed but a little faster than normal.

16 VCR Servo Problems: The Diagnostic Device Revisited—Part 1

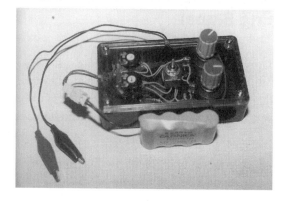

Figure 16-3. This is a photo of the diagnostic device I constructed. I used a 4.8V Ni-Cd cordless telephone battery to power it.

I adjusted the fine adjust control until the sound seemed normal. The picture would alternate between clear video and snow as the video heads wandered into the guard bands and wrong azimuth tracks. This is normal for a properly working capstan motor when being manually controlled by an adjustable voltage source. Best results are obtained when using a tape recorded in the SP mode for this test. In other modes, the capstan motor runs so close to its stall point that it may need frequent restarting.

Figure 16-4. I opened the capstan motor control line and connected the device's positive lead to the motor side of the line and the negative lead to the VCR's ground.

Examining the Synchronizing Pulses

Again I observed the FG pulses at the FG head. Their amplitude and frequency now were almost constant, but without the schematic I couldn't follow them through their conditioning circuits to the servo IC. What's more, I could find no pulse at any pin of the servo IC that had the characteristic trapezoidal appearance and frequency associated with the fully conditioned FG pulse.

VCR Servo Problems: The Diagnostic Device Revisited—Part 1   16

However, I was able to find the drum FG and PG pulses, which looked normal. The FG pulse frequency was approximately 719.3Hz and the PG pulse frequency was 29.97Hz. To verify that these pulses were drum related, I loaded the drum motor by lightly pressing my finger against its upper surface. The frequencies of the two pulse trains changed.

Measurement showed that the line that normally provides the control voltage to the motor was stable when the motor was running at a constant speed. However, the voltage on that line would increase to about 7V when I lowered the motor speed to a certain point, and decrease to about 0.1V when I raised the speed.

The servo circuits were making some attempt to control the speed of the motor, but the control was far too loose. In any event, it looked as though the erratic FG pulses were somehow related to the erratic motor speed control.

Obtaining a Schematic

At this point I decided that I was going to have to either obtain the schematic or resort to the "shotgun" method. Since shotgunning has no educational value, I opted to get a copy of the schematic. Understanding the exact cause of this problem was more important to me than trying to maximize profit.

When the schematic arrived (*Figure 16-5*), I focused my attention on the path from the FG head to the pin labeled "C.FG" on the TD6360N-02 servo control IC. The sine wave signal that was induced into the FG head by the rotating capstan magnets appeared normal. It was routed to the base of Q207, which was a preamp in a common emitter configuration.

I observed the signal at the base of that transistor with the diagnostic device connected and the VCR running at constant speed. I found the pulse amplitude to be 30mV$_{pp}$. The signal at the collector was 700mV$_{pp}$, showing that the stage provided considerable gain.

The signal was routed from the preamp to pin 2 of IC203, which was a KIA75902P quad op amp. Pin 2 was the inverting input of the amp labeled "C.FG AMP". This amp is configured in such a way as to be driven alternately between saturation and cutoff by the input signal. This causes clipping of the top and bottom of the sine wave, creating a quasi-trapezoidal wave. Though the input to this amp looked normal, the output resembled a very narrow positive-going spike.

16 VCR Servo Problems: The Diagnostic Device Revisited—Part 1

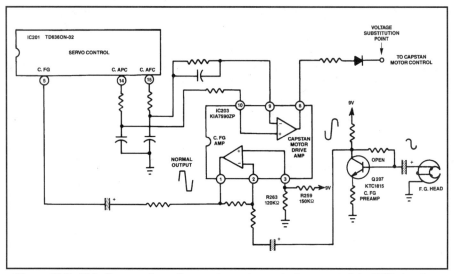

Figure 16-5. This diagram shows the capstan motor control segment of the VCR circuits. Note where I substituted the voltage from the diagnostic device.

Solving the Problem

Because IC203 also contained the properly working drum motor amp, I surmised that it was good, although it is not uncommon for a single amp in a multi-amp chip to fail. The non-inverting input (pin 3) of the capstan FG amp should be held at close to one-half of the 9V supply by the voltage divider consisting of R263 and R259. Measurement showed about 0.1V at this pin. A quick check of these resistors showed that R259 (150KΩ, 1/8W) was open. This caused the non-inverting input to be pulled close to ground by R263.

The output of an op amp swings high when the noninverting (+) input is more positive than the inverting (-) input and low when the situation is reversed. With the (+) input held at ground because of the open resistor and the (-) input above ground the op amp's output would always be low. When the amplitude of the sine wave became great enough, the negative peak would force the (-) input below the (+), which would drive the output high.

Scoping the input and the output of the op amp simultaneously showed that the positive spike occurred during the interval of the negative peak (*Figure 16-6*). The spike was only present when the motor was running well above its normal speed. After replacing R259 I could see a clean trapezoidal wave at the op amp's output regardless of the motor's speed (*Figure 16-7*). The amplitude of

VCR Servo Problems: The Diagnostic Device Revisited—Part 1   16

the wave was about 7.5V$_{pp}$. I disconnected the diagnostic device and reconnected the motor control line. The machine ran perfectly. So exactly what had been happening?

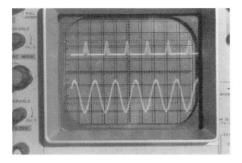

Figure 16-6. This oscilloscope photo shows the FG signal at pin 2 of IC203 (sinewave at bottom), a quad op amp. Pin 2 was the inverting input of the amp labeled C.FG AMP. The spike at the top of the screen is the output of the op amp. The positive spike, present only when the motor was running well above its normal speed, occurred during the interval of the negative peak.

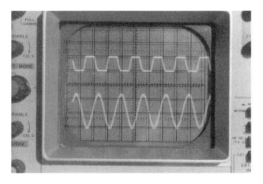

Figure 16-7. After replacing R259 I could see a clean trapezoidal wave (the correct signal) at the op amps output regardless of the motor's speed.

The Problem Solved

The AFC circuit within the speed control loop responded to the absence of the FG pulses by causing the motor speed to increase. This is understandable because an absence of pulses corresponds to the lowest possible speed (stop), which calls for the maximum correction. The fact that the motor was turning made no difference. Since the amplitude of the pulses induced in the FG head are proportional to the motor speed, the higher motor speed resulted in higher-amplitude pulses.

When the pulses reached a level where they could cause the improperly biased op amp to output a signal that was detectable by the AFC block, their frequency was too high. The servo circuits responded to this by instantly slowing the motor, which subsequently reduced the amplitude of the FG pulses below the level of detection. The result is the continuous changing of the motor's speed.

16 VCR Servo Problems: The Diagnostic Device Revisited—Part 1

Servo circuits aren't difficult to repair if you have a good understanding of how they work and a basic knowledge of electronics. Using the diagnostic device, or another suitable voltage source, to open the loop would be extremely helpful in localizing the problem. Once this is done, normal troubleshooting procedures will lead you to the defective component.

Chapter 17
VCR Servo Problems: The Diagnostic Device Revisited— Part 2
By Steve Babbert

Part one of this article on troubleshooting VCR servo systems, which appeared in the September issue of ES&T, focused on the capstan speed and phase loops. This part will examine the role that the servo circuit plays in maintaining the correct speed and phase of the head drum or "cylinder" motor. This discussion will be based on the circuitry used in a Magnavox Model VR9547AT01. After a brief circuit description we will look at a specific problem. As in part one, the diagnostic device will be used to open the loop and manually control the motor's speed.

The Servo Circuit

In this chassis, most of the circuitry associated with the capstan and drum servo is contained in IC 2001, an MN6178VAD capstan/cylinder servo processor. Two other ICs used are IC2003, an AN3793 cylinder servo interface and IC2603, an AN3812K cylinder motor driver (*Figure 17-1*).

In this system the FG and PG pulses that are used as feedback signals are both generated by a single Hall-effect device, which is part of the drum assembly. The positive and negative outputs of the Hall device are routed to pins 8 and 9 respectively of the cylinder motor drive IC. Inside this IC these outputs are applied to the PG/FG generator block. The output from this block at pin 15 is a composite signal consisting of the amplified and shaped pulses.

At this point the PG pulse has a frequency of 29.97Hz and an amplitude of about $4.6V_{pp}$. The FG pulse frequency is 179.8Hz with an amplitude of about $2.4V_{pp}$. Viewing this pulse train on an oscilloscope shows five FG pulses followed by one PG pulse per cycle (*Figure 17-2*). This pulse train is applied to pin 17 of the cylinder servo interface IC where it will be routed to the PG/FG separator block.

17 VCR Servo Problems: The Diagnostic Device Revisited—Part 2

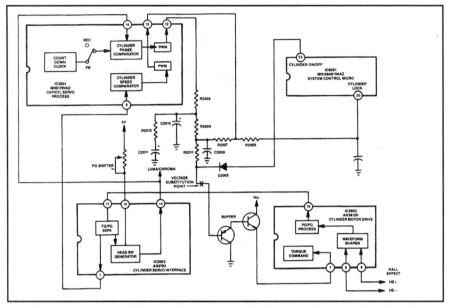

Figure 17-1. Main circuitry associated with the cylinder servo of a Magnavox Model VR9547AT01 VCR.

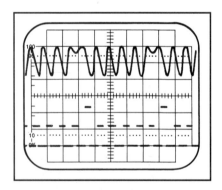

Figure 17-2. The top waveform here is the output of the Hall Device, observed at pin 9 of IC2603. The lower waveform is the output of the PG/FG processor at pin 15 of IC2603.

The PG/FG Separator

The separator has two outputs. The FG output exits at pin 1 and enters pin 9 of the capstan/cylinder servo process IC. It is used for comparison in the speed comparator block.

The PG pulse is applied to the head switch generator. The output of this block is a 50/50 duty cycle square wave that shares the frequency of the PG pulse

VCR Servo Problems: The Diagnostic Device Revisited—Part 2   17

(*Figure 17-3*). The phase of this signal can be retarded or advanced with respect to the incoming PG pulse through adjustment of the PG shifter control connected to pin 15.

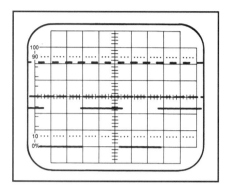

Figure 17-3. *The top waveform is the separated PG pulse output at pin 1 of IC2003. The lower waveform is the separated and stretched PG pulse at pin 14 of IC2003 used for headswitching and phase comparison.*

The output of the head switch generator follows two paths. One path leads to the luma/chroma circuits where it will be used for headswitching. The other path leads to pin 14 of the servo process IC where it will be used for comparison by the phase comparison block. A 3.58MHz signal is also applied to this IC where it will be divided to produce the stable comparison references.

Error Voltages

Error voltages are generated by the speed and phase comparators in response to differences in the frequency and phase of their input signals. These voltages are applied to their respective PWM blocks to control their duty cycles.

The outputs of the two PWM blocks, after being low-pass filtered and combined by a passive RC network, are routed to pin 7 of the cylinder motor driver IC. In this IC the control voltage is applied to the torque control block. This block coordinates the three phase voltages that are applied to the windings of the direct drive motor.

Solving a Servo Problem

When the VCR was placed into the play mode and the tape was fully loaded, I could hear what sounded like a motor speeding up then slowing down at a frequency of about once per second. The video on the monitor was pulling to the left and flagging at the top in step with the speed variations. The sound

17 VCR Servo Problems: The Diagnostic Device Revisited—Part 2

from the linear sound track was normal. Normal sound from a linear track usually indicates that the capstan motor speed is okay. For this reason, I focused my attention on the cylinder.

Though the motor sounded like its speed was fluctuating, I couldn't see any change. Before I began troubleshooting I wanted to be certain. There are several ways to determine the speed of a cylinder motor. One way is to monitor one of the feedback pulses. Their frequencies are proportional to the motor's speed. Another way is to use the strobe method.

The Strobe Method

If a spot is placed on the upper surface of the cylinder toward the outer edge, it will produce a characteristic pattern when viewed spinning under a fluorescent light. A fluorescent light has a 120Hz flicker rate when powered by a 60Hz voltage. The rotational speed of the cylinder is close to 30 revolutions per second. This results in illumination of the spot four times per revolution.

The pattern produced by this stroboscopic effect on the spot on the cylinder looks like a ring with four gaps spaced at 90-degree intervals. If the cylinder speed was exactly 30rps the pattern would appear stationary. Since it is 29.97rps, the pattern rotates slowly clockwise. It takes approximately 36 seconds for the pattern to make one revolution in a normally working machine.

When I perform this test I place a piece of tissue paper in one of the holes near the outer edge of the cylinder's upper surface. After saturating the paper with the ink from a pink fluorescent highlighter, I view the rotating cylinder under an ultraviolet (black light) lamp. This makes the pattern very bright.

In this case, the pattern reversed direction from clockwise to counterclockwise in step with the speed variations that I could hear. This verified that the cylinder speed was unstable.

Scoping the PWM Outputs

With the vertical inputs of the scope set for dc, I simultaneously scoped the phase PWM output (pin 12) and the speed PWM output (pin 13) of IC 2002. During the acceleration burst there was a square wave present at the phase PWM output. Between bursts, the output became a steady line.

VCR Servo Problems: The Diagnostic Device Revisited—Part 2 17

Counting graticules showed that this line represented a 5V output. At the speed PWM output a squarewave was always present with a duty cycle that increased during acceleration of the cylinder, and decreased during deceleration (*Figure 17-4*). Because R2011 couples the low-pass filtered control voltage to the motor driver it was a good place to open the loop.

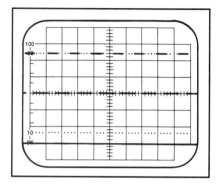

Figure 17-4. The upper waveform is the speed PWM output (pin 13). The duty cycle of this waveform varied during acceleration bursts. The lower waveform is the phase PWM output at pin 12. This waveform showed 5V between acceleration bursts.

Opening the Loop

After wicking the solder from one leg of R2011, I lifted it from the board. This completely isolated the low-pass filter from the motor control line labeled "cylinder error." This line leads to a two transistor buffer amplifier before being applied to pin 4 of the cylinder motor drive IC. By taking control of the line at this point I could test the buffer amp, the motor drive control circuits, and the motor.

Using the Diagnostic Device

Because I couldn't find a convenient place to tie into the cylinder error line, I tack soldered a jumper to the land at the cathode of D2005 (*Figure 17-5*). The anode of this diode is connected to pin 53 of the system control IC6001. In any mode other than play or record this pin goes high, forward biasing the diode and forcing the cylinder motor to stop.

Next I connected the positive lead of the diagnostic device to the jumper and the negative lead to ground. While still scoping the two PWM outputs of IC 2002 and watching the video monitor, I placed the VCR into the play mode.

17 VCR Servo Problems: The Diagnostic Device Revisited—Part 2

This VCR model uses "negative logic" in which a lower dc output from the PWM blocks causes the motor to speed up and a higher voltage causes it to slow down. This must be considered when using voltage substitution. Once the tape was fully loaded I used the reverse kick-start function on the diagnostic device. This lowers the control voltage causing the cylinder motor to begin running at a higher than normal speed. I then returned the switch to the run position and began to adjust the coarse control while watching the speed PWM output.

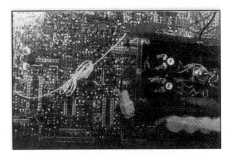

Figure 17-5. To break the servo loop and control the motor manually, used desoldering braid to remove the solder from one leg of R2011 and lifted it from the board. This completely isolated the low-pass filter from the motor control line labeled "cylinder error." I then tack-soldered the diagnostic device to the land at the cathode of D2005, the input to the buffer amp.

Adjusting the Cylinder Motor Speed

As the speed of the motor became close to normal, the PWM output approached a 50/50 duty cycle. At this point the video on the monitor locked in. I found that through careful adjustment I could make the video lock in for 15 to 20 seconds at a time. This is usually not possible when controlling a capstan motor.

The cylinder motor is inherently more stable than the capstan motor because of its greater inertial mass. This is one reason why the video can be locked in for longer periods. On the other hand, this leads to a more sluggish response. When you adjust the controls on the diagnostic device you must allow time for the motor to come up or down to speed. After some practice it becomes easy.

Scoping the phase PWM output showed that it would begin toggling between high and low as the speed was changed. Since the scope was set for dc, the trace would shift up and down between 0V and 5V. When the motor was brought close to the correct speed, I saw that just prior to these transitions the flat line would break into a square wave. The duty cycle would increase or decrease depending on whether the transition was from low to high or vice versa.

VCR Servo Problems: The Diagnostic Device Revisited—Part 2 17

If I made careful adjustment, I could get the square wave to lock in for a few seconds, though its duty cycle wavered. The phase PWM output is normally very touchy when making this test so this section was not suspect.

If I raised the speed of the cylinder motor much above normal, the output of the phase PWM would default to a 50/50 duty cycle. At the same time, the duty cycle of the speed PWM output would become 100%, which results in the 5V flat line (*Figure 17-6*). When I slowed the speed of the motor to the point where the duty cycle of the speed PWM was about 20%, the machine would shut down. This shut-down function will be discussed shortly. In the shut-down mode the phase PWM output defaults to a 50/50 duty cycle and the speed PWM output will show a zero volt flat line.

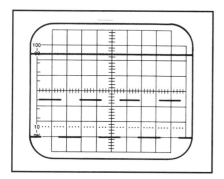

Figure 17-6. The upper waveform shows the speed PWM output (pin 13) at 100% duty cycle with the motor speed raised above normal. The lower waveform is the phase PWM output (pin 12). This signal defaults to a 50-50 duty cycle when the motor speed is raised above normal.

What the Tests Showed

So far the tests had shown that the cylinder motor and its drive circuits were functioning normally. The PWM blocks seemed normal too. From this it followed that the comparators that drive the PWM blocks were normal. Since the error amps require feedback from the motor in the form of FG and PG pulses, apparently the feedback was present and correct. This ruled out the cylinder servo interface IC, and any sections of the motor drive IC associated with feedback and the Hall device. The only section left untested in the loop was the low-pass filter.

The Low-pass Filter

The speed PWM output is low-pass filtered by an RC network consisting of C2009 and R2007. This output is also routed via R2006 to the "cylinder lock"

17   VCR Servo Problems: The Diagnostic Device Revisited—Part 2

(pin 20) of the system control microprocessor IC6001. If the cylinder speed becomes too low for any reason, the resulting lower duty cycle with its lower average dc value will pull pin 20 of IC6001 low. The system control responds to this by forcing the machine into the stop mode. The normal voltage on this pin is about 3.2V. When it drops to about 2V, shutdown will occur.

The phase PWM output is low-pass filtered by the RC network consisting of R2008, R2009, C2010, C2011 and R2010. The two capacitors in this filter are electrolytics. I checked these first, because electrolytics are prone to failure. A plug-in circuit board assembly which held most of the voltage synthesizer circuitry had to be removed to gain access. Capacitor C2011, a 6V 100µF unit tested bad. After I replaced this capacitor and reconnected R2011, the machine ran normally.

Analysis of the Problem

When the low-pass filter fails, the output of the PWM blocks won't be smoothed to the dc value. As the cylinder drive circuits attempt to follow the ac component, the motor's speed will become unstable. This causes the FG and PG pulse frequencies to fluctuate, which creates additional instability. The sluggish response of the motor limits the rate at which its speed can fluctuate.

Whenever I complete a repair, I make it a habit to take a second look at all pertinent waveforms and voltages taking notes as needed. Observing various parameters of a normally working machine will be helpful when trying to spot abnormalities in the future. Natural curiosity, a willingness to snoop and a good understanding of electronics lead to proficiency.

With the VCR in the play mode I scoped the outputs of the two PWM blocks. There were clean square waves at both pins with duty cycles of about 50% (*Figure 17-7*). The frequencies of the speed and phase PWM outputs were about 7KHz and 3.5KHz respectively. Loading the cylinder motor by applying light pressure with my finger resulted in a reduced duty cycle of both of these signals.

This concludes part two of this series on troubleshooting VCR servo systems by using voltage substitution. Using the diagnostic device or another suitable voltage source to open the loop will save time and enable you to see cause and effect relationships.

VCR Servo Problems: The Diagnostic Device Revisited—Part 2 17

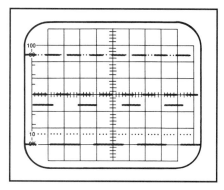

Figure 17-7. These waveforms show the speed PWM (top) and the phase PWM (bottom) outputs after the problem was corrected.

Chapter 18
Where Do I Begin?: Analyzing VCR and Camcorder Problems
By Steven Jay Babbert

When preparing to service a VCR, my most valuable service information often comes from the customer. Typical questions I ask are: "What is it doing—or not doing?" "Did the problem appear suddenly or gradually?" "Was the unit handled improperly or abused in any way?" Answers to these questions can be helpful in localizing the problem even before the cover is removed.

A fairly common complaint is that the VCR starts to play but then stops and possibly ejects the tape or shuts down. A symptom such as this could be caused by a variety of problems ranging from a worn belt or idler wheel to a defective syscon (system control microprocessor). Information provided by the owner could be helpful in this case.

If the owner tells me that the unit has had a fast-forward or rewind problem that has been getting progressively worse, I'll focus my attention on parts of the drive section that are common to the play, FF and rewind modes. The idler wheel is subject to progressive wear and may not cause play-mode problems until it becomes seriously worn.

If the unit has had no prior problems the drive system could still be at fault particularly in the case where a belt has broken or fallen off, but now there is an increasing likelihood that a non-mechanical problem exists. In this case visual inspection will be helpful while trying to cycle the unit through various modes. This should be done while using a clear or cutaway test jig.

The Test Jig

The cassette test jig is a valuable troubleshooting aid, particularly for tape transport problems. Not only does it prevent possible damage to expensive test tapes, it allows you to see moving parts which would otherwise be blocked. More than one test jig can be useful since different models allow access to different areas. Be sure that the jig is designed for the chassis you are servicing. For

18 Where Do I Begin?: Analyzing VCR and Camcorder Problems

example, "G" function chassis requires a "G" function jig. These jigs will also work on most other models.

Test jigs perform two main functions; they activate the respective "tape-in" leaf switches, informing the syscon that a tape has been installed, and they prevent light from the IR LED from reaching the tape-end sensors. Some older models use miniature incandescent lamps instead of LEDs.

In normal operation, clear tape leaders allow light to pass, signaling the syscon that the tape has come to an end. The syscon responds by stopping the drive motor and applying the brakes to prevent damage to the videotape.

In some VCRs, ambient light will activate the photo sensors when a test jig is used with the cover removed, causing the machine to stop. Usually this can be remedied by blocking light from specific areas using cards or by using lower light levels.

As a last resort you can cover the photodetector openings with black plastic tape. These will be located near the cassette door hinge on either side of the tape stage while in the fully-loaded position. If you must cover the photo sensors be sure to remove the tape after servicing.

Generally when using a test jig in any mode, the VCR will stop after a few seconds. This is because the feed reel won't be turning without an actual tape. This condition is sensed as a fault by the syscon which initiates shutdown. Turning the reel by hand will prevent this problem. It doesn't usually matter how fast or in which direction.

Reel Motion Sensor

Reel motion is typically sensed by an IR LED/photo detector assembly located beneath the reel table. If the LED and detector are housed in the same package it will most likely have four wires. The underside of the reel tables have a series of mirrors that alternately reflect or block reflection between the LED and detector as the reel rotates. The resulting pulsed signal generated by the turning reel is applied to the syscon.

In some units the tape counter is incremented by the reel motion sensor pulse and may be used to indicate the presence of pulses. This does not apply to VCRs that have "real-time" counters.

The take-up reel should be turning once the machine has settled into the play mode. With practice you will be able to make a judgement as to the general condition of the drive system by grasping the reel with your fingers and feeling for torque. Use a torque gauge if you're in doubt. Bear in mind that if you inhibit the reel motion for very long the machine will stop. I usually allow the reel to turn at a reduced speed while trying to feel for any loss of power.

If the take-up reel torque is low or intermittent the test jig should allow you to see where the power is being lost. In most cases the idler tire will be the culprit. It might appear cracked or glazed. In some instances cleaning may be all that's necessary to remedy the problem but usually this is only a temporary fix.

In most designs the idler wheel is driven indirectly by the capstan motor. Other parts of the idler assembly can break down causing loss of torque. The reel tables themselves contain a slip-clutch mechanism consisting of felt pads, springs and pressure plates. These can be rebuilt if they appear to be worn. Be sure to replace any bushings after reinstalling a reel table as height is critical.

Watch the Take-up Reel

Another method of checking the tape transport system if you suspect it is causing a shutdown problem, is to focus your attention on the take-up reel while playing a standard tape. If the take-up reel is turning smoothly but the VCR still stops or shuts down consistently after a few seconds, then the problem must not be related to the idler assembly or other parts of the drive section. In this case I would check the reel motion sensors and associated circuitry.

If the take-up reel turns intermittently or freezes up just before shutdown, then the transport system must be at fault. In addition to a slipping idler wheel, this type of problem could be caused by binding at some point in the tape path or binding of the feed reel. The feed reel has brakes and a brake band to provide quick stopping, and drag to create the proper tape tension.

If you suspect excessive tape tension you might need to use a tape tension gauge. However, in many cases visual inspection may help you to determine that a brake isn't being released due to a mechanical failure. Insufficient tape tension can lead to loss of video, tracking or sound due to improper contact between the heads and the tape.

18 *Where Do I Begin?: Analyzing VCR and Camcorder Problems*

Idler Replacement

In the event that the idler wheel has been found to be worn, there are two choices: you can replace the entire idler assembly or you can replace just the idler tire. Some techs routinely replace the entire assembly. In this case a higher bench charge can usually be justified. I rarely replace more than the tires, which I can usually find in a universal tire kit. This policy saves my customers money and in turn they become my best advertisers. The decision whether to replace just the tire or the entire assembly, should, however, be made on the basis of the condition of the idler assembly as a whole.

The pinch-roller and capstan regulates the speed of the tape being drawn past the various heads and guides. In most cases the speed of the capstan will remain correct even when problems develop in the idler/take-up system. In fact, this is one cause of tape "eating."

The take-up reel may lose torque and fail to pull in the tape fast enough and yet continue turning just enough to avoid system shutdown. Since the capstan is still pulling tape at a constant speed the excess will spill into the machine.

Another tape eating problem occurs during unloading with a worn idler. During loading the tape is pulled from the cassette by guideposts on sliding tracks. During unloading the guideposts retract but the take-up or feed reels must pull in the tape slack. An idler problem can prevent this from happening and when the cassette is ejected a tape loop will be caught in the machine or at least be left hanging out of the cassette.

Pinch Roller Problems

If the pinch roller becomes glazed it may lose its grip on the tape. The take-up reel tries to turn faster than necessary to maintain tape tension while the slip-clutch keeps the tension in check. In this situation the tape may move faster than normal as it slips past the pinch-roller or it might stop moving depending on the combination of torque and drag in a given machine. Replacement of the pinch roller is the best solution in this case though I have resurfaced hard-to-find pinch-rollers with a hobby grinding tool and a fine sanding bit.

Where Do I Begin?: Analyzing VCR and Camcorder Problems **18**

System Control Problems

Like most mechanical functions, loading is coordinated by the syscon. A leaf switch is closed when a tape is inserted into the VCR which instructs the syscon to begin the load sequence. Loading problems can result from an electronic malfunction but most of the time the problems are mechanical.

In addition to the tape-in switch which initiates loading, a multi-position "mode" switch allows the syscon to monitor various mechanical movements so that it can activate or deactivate motors and other components at the appropriate time. This switch is mechanically linked to the tape "stage" (the tape holding compartment) and other moving parts. A problem with this switch or its synchronization can prevent loading or playing after loading.

A common loading problem results from foreign objects in the machine. When a tape is inserted, the loading sequence will begin only to be halted when contact is made with the object. In most cases the tape will be ejected after a few seconds because the syscon hasn't received verification from the mode switch that loading has been completed. Without this safely feature the load motor would burn out eventually by continually trying to load the tape.

The tape-in switch can easily be checked with a DMM. In most cases you will be able to see if either of the leaves has become misshapen preventing proper closure. The mode switch on the other hand isn't so easy to test because there are so many configurations. They typically have about five connections. Service literature will usually include a table showing where you should expect to find continuity in each position.

If you suspect that a mode switch is faulty, particularly in a case of intermittent operation, try applying pressure to various parts of the switch assembly while cycling the VCR through the mode in question. In some cases the position of the mode switch or part of the actuating linkage is adjustable and might have shifted out of position.

Occasionally you might service a unit in which the loading gears have been forced out of sync by someone trying to force loose a tape which had become lodged. The tape hang-up might have been the result of a problem within the cassette itself which prevented ejection and caused the machine to shut down. In these cases minor problems have been turned into major ones. If you don't have service literature a known good unit of the same design can be used for comparison.

18 *Where Do I Begin?: Analyzing VCR and Camcorder Problems*

When the Unit Blows Fuses

Blown fuses often indicate a serious overload caused by a shorted or leaky component. The problem could be in the power supply or in one of the circuits it supplies. In some VCRs the power supply is located on a separate subchassis. In some models a plug-in harness can be disconnected to help isolate the problem. Service literature might be required for this type of problem but searching for low resistance readings to ground may lead you to the defective component.

Fuse blowing symptoms aren't necessarily confined to electronic problems. In one VCR the fuse would blow only when the unit was placed into play, FF or rewind. At first I suspected a shorted motor or motor driver IC. To my surprise I found that a stretched belt had become lodged between the capstan motor flywheel and the board, seizing the capstan motor. When power was applied to the motor it began drawing excessive current because it had an excessive load. Apparently this was enough to blow the fuse.

Another place objects become lodged is in one of the tracks used by the loading posts that pull the tape around the head drum. If these posts aren't firmly seated against the "V-stops" at the end of their tracks, part or all of the picture may be missing. In some cases the lodged objects will be from the VCR itself. Screws and retainers that have worked loose have a habit of finding their way into this area.

Motor Control Circuits

The motor control or "servo" circuits in VCRs are very complex. These circuits maintain the correct speed and phase of both the head drum and capstan motors. Some servo problems will cause symptoms similar to those of defects in other circuits. No video with only snow (some models go to a blue screen under these conditions) may look like a head problem when in fact the heads aren't tracking due to a servo problem. Past issues of this magazine have featured articles devoted to solving servo problems so they won't be covered in depth here.

If the sound or video is playing at the wrong speed, particularly if the speed is erratic, there is a good chance that the capstan servo circuits are faulty. In some cases it will appear that the servos are hunting for the correct speed as the capstan motor alternately speeds up and slows down. In most cases this type of symptom is not due to motor problems.

Problems with the head drum speed or phase servo won't affect sound recorded on the linear sound track which is picked up by the stationary audio head. Video might drift in and out, noise bars might be present or video might be completely absent. Bad bearings in the drum motor can also cause some of these symptoms. Sometimes turning the head drum by hand and feeling for play or catching will help reveal this type of problem.

One way to differentiate between servo problems and problems in the motor or motor driver circuit is to use voltage substitution. This works because the servo circuits use a dc voltage to control the motor speed. However, the dc voltage is applied to the motor driver circuit and not the motor itself. These motors are usually three-phase and cannot be directly tested like the loading motor with dc voltage.

Mechanical Alignment

It is usually necessary to follow the manufacturer's recommendations when doing mechanical alignments. Most adjustments require special tools which can be purchased in sets. Fortunately alignments are rarely needed except during replacement of parts in the tape path.

Gross misalignment can be spotted by watching how the tape rides against the guideposts. If the tape rides too high or too low it will "bow" or even ride outside of its track. This will affect the way that the tape meets the video heads and can cause problems in part or all of the picture.

Misalignment of the audio/control head assembly can cause problems ranging from poor sound to servo system failure. Since the servo system requires feedback from the control head it is essential that it be tracking properly. The audio/control head assembly generally has at least two adjustments.

If you suspect an alignment problem but have no success using the recommended procedure, consider the possibility that one of the reel tables is at the improper height. If a reel was removed during repair and a bushing was lost, the height could be improper. Reel height gauges are available but specific models need specific gauges.

18 Where Do I Begin?: Analyzing VCR and Camcorder Problems

The alignment of the "V-stops" is very critical and requires a special test jig only available from the manufacturer. These rarely need adjustment even if guideposts have been replaced. Don't overlook the possibility that they might have shifted particularly if untrained personnel have been involved.

The rf Modulator

The main function of the rf modulator is to convert the baseband audio and video signals into an rf channel which is switch selectable to channel 3 or 4. This signal is available at the output F-con when the VCR/TV mode is set to VCR. The modulator housing also holds the input F-con that brings the input signal into the VCR's tuner. This input signal is also passed to the output F-con when the TV mode is selected.

A simple test for the rf modulator is to check for baseband audio and video at the line-level A/V output jacks. These signals can be scoped or fed to the A/V input of another VCR or a monitor. If they are present during tape playback, or when a channel is selected on the tuner, it is likely that the rf modulator is defective or not getting power. These are usually not difficult to obtain or replace.

RF modulators can develop other problems besides complete failure. They can develop reduced power level causing a weak picture or they can introduce noise or cause the picture to become grainy. Whatever the symptom, if it is present during both tape playback and while using the VCR's tuner then the problem is probably in the modulator.

Detached F-con

Occasionally you will come across an RF modulator in which one of the F-cons has been torn loose. Usually this is the result of a dropped cable converter etc. In most cases I've found that the modulator can be repaired. Sometimes the original F-con can even be reinstalled.

After desoldering and removing the modulator, remove the rear shield (consider the F-con side as the front). If the entire F-con was pulled out you will see a hole usually with a metal grommet in the circuit board where the center conductor was soldered in place. If the center conductor is still in place you might be able to desolder it and pull it through from the rear, otherwise you might need to further disassemble the modulator.

Since the original F-con was probably pressed into the front housing, the housing might be slightly deformed. Flatten the deformed area if necessary. It won't affect the performance of the modulator, but in some cases the modulators can't be reinstalled properly unless the face is flat.

The new F-con will have to be soldered to the housing. This might require a hotter iron since the heat tends to dissipate through the housing. If you're using a temperature controlled iron you might be okay since they tend to hold the tip temperature well.

First install the F-con and solder the center conductor into place. Next solder the F-con to the housing one point at a time. It is a good idea to follow this step by cooling the F-con with circuit coolant. If the plastic insulator in the F-con reaches its melting point it will quickly deform and be ruined.

Since most of the work involved has to do with removing and reinstalling the RF modulator, I have found that the bench charge is about the same as when replacing the modulator completely. I have, however, saved my customer the cost of a new modulator. I find that in most cases they are very pleased by this which is more important to me than what little extra profit I could have made through mark-up of a new modulator.

The Vacuum Fluorescent Display

Most VCRs use a vacuum fluorescent display which requires a supply voltage somewhat higher than the typical voltages used by other circuits and devices. In some models these voltages are generated by the power supply. Other models use a dc to dc converter module which builds the higher voltage from a lower one. If a "dead" display is found in an otherwise working unit, check the converter first. If the input voltage is good but the output voltage is low or missing it is probably defective.

18 *Where Do I Begin?: Analyzing VCR and Camcorder Problems*

Remote Control Problems

Remote control problems are relatively uncommon but they do occur. When a VCR responds to front panel controls but won't respond to the remote control, the first thing to do is isolate the problem to the sending or receiving unit. The sending unit can be tested with an IR sensitive strip or card that converts the IR to visible light. These are available through many parts houses. Just point the remote control at the strip and press any button. You will be able to see a flashing red light on the sensitized area. These can also be used to check the IR LED which is used for the tape end sensor.

If the problem isn't in the sending unit it could be anywhere between the VCR's IR detector and the syscon. In many VCRs the receive unit is self-contained having an IR detector, amplifier and waveshaping circuits. The output is coupled to the syscon which decodes the information and gives the appropriate response. Scoping the line between the IR module's output and the syscon's remote input may help to further isolate the problem. Models not using a self-contained module will have an IC between the detector and syscon.

Syscon

If a system control (syscon) problem is suspected it may be necessary to obtain a schematic to identify the various pins. In order to operate, the syscon needs a supply voltage (typically 5V), a clock signal and a reset pulse to initiate the internal program when the VCR is powered up. The clock is usually part of the syscon and only an external crystal is required to set the frequency.

If the supply voltage, reset pulse and clock signal are present then the syscon should be running. Logic-level high or low transitions at the various input pins from sensors or switches should be followed by transitions at the appropriate output pins. If not the IC must be suspected. Since these are usually static-sensitive CMOS devices, be sure to observe proper handling precautions.

Cleaning and Lubrication

Cleaning and lubrication is often all that's needed to restore a VCR to proper operation. The proper materials and common sense are all that's required. Be especially careful when cleaning the fragile video heads. Use a gentle side-to-side motion with chamois swabs and always use fluid designed for the job.

Regular lintless swabs are suitable for audio/control and full-erase heads. Whenever possible I try to clean the idlers and belts while the machine is running with a cassette test jig by holding the swab against the moving plastic or rubber. Be sure to clean all guides that come into contact with the tape. Use a dry swab to remove fluid with dissolved contaminants.

I generally lubricate any unit that hasn't been serviced for a year. As a rule-of-thumb, oil things that turn, and grease things that slide or mesh. Use only the recommended lubricants and be careful how you apply them! Oil on belts, tires or felt pads can cause a call-back.

Summary

VCR repairs aren't too difficult once you develop a basic understanding of the system as a whole. Diagnosing problems might be challenging at first but confidence will grow with each successful repair. In time you will familiarize yourself with "trouble spots" in specific models and learn to track down problems by using symptom analysis.

Chapter 19
VCR Service Centers: Tips on Remaining Profitable
By Wayne B. Graham

Making money in the electronics servicing business has required many changes since the TV repair/tube tester days of the 50's, 60's and early 70's. Even the early days of $750 to $1,000 VCRs were good for the electronics service business. Repair quotes of $75 or $150 were thought to be a good deal, and these older machines were designed to be repaired; so, it wasn't a frustrating experience for the technician. But now many consumers are thinking of the VCR as a disposable appliance, like an alarm clock or telephone.

The Arithmetic of VCR Servicing

Are service centers still able to make money servicing VCRs? Yes, but it is a lot tougher these days and many service centers won't survive the transition caused by the low-priced new units available on the market. We've talked with thousands of servicers over the years, and it's been interesting to note the differences between service centers that make money, grow, and stay in business, versus those that don't make money, dwindle, and end up with disconnected phone lines leaving "no new number."

From the thousands of technicians and service center owners we've talked with over the past few years, here's our outline for a profitable VCR service business.

Charge For Estimates

Charge $20 to $30, depending upon the area of the country and local cost of living. If you do free estimates, the customer may continue to shop around for the lowest repair cost and the machine may not make it back to your service center, even if you were the low bidder.

Unfortunately, those who don't charge for estimates often don't have a very well equipped service area, and you may lose out to someone with not much more test equipment than a multimeter and needle nose pliers. Charging for the estimate also provides a source of repairable VCRs, since the customer often chooses to abandon his VCR instead of going ahead with repairs. You either get the estimate fee, the repair job, or the VCR, so at least you aren't working for free.

Provide a Knowledgeable Accurate Estimate

You might call this a "technical evaluation" rather than an estimate, because that's really what it is. The fact is the mechanical parts of the VCR do wear out. If you review any factory service manual, you'll see that such mechanical parameters as hold back tape tension, take-up torque, FF torque, rewind torque, reel table heights, guide heights and other similar measurements represent the basics for mechanical performance of a transport.

If you're using a "trial and error" method, it will prove to be too prone to errors and too time consuming. Often there are several marginal problems along with the obvious problem. Conscientious as you may think you are, the guessing that is necessary without proper test equipment will help assure failure.

Test equipment will allow accurate, true estimates in the least amount of time, plus this same test equipment can be used to help perform the actual repairs.

The ability to measure the actual amount of video head tip protrusion (wear) has proven to be one of the most important parameters for determining not only how much use the machine has already had, but more importantly how much longer it's likely to last. Accurately knowing video head wear allows you to honestly advise a potential customer that; "even though the machine is six (or whatever) years old, the heads are still in very good shape and should last for six (or whatever) more years, once these 'other' problems are corrected."

Having information like this helps assure the customer that you know what you're doing and that his machine will be fixed correctly by an expert. We recommend the use of a check out form to compile findings regarding these critical electronic and mechanical tests. The form provides an excellent method of showing expertise to the customer, and helps to fully explain and justify repair estimates. It can also be maintained as a history as future machine problems arise. This leads us directly to the warranty on your work.

Provide a Six-month To One-year Warranty on VCRs You Service

I'm always amazed at how many technicians tell me that they only offer a 30-day warranty, just on their work, but then complain about lack of business. Should it really be a surprise? New machines offer a one-year warranty, even longer when purchased on a number of different major gold credit cards. The customer has to weigh the cost of $150 for a new machine with a one or two year warranty versus $65 (national average) for a "how many day warranty?"

If you are able to quickly and accurately measure all of the normal wear items in a transport, including video heads, tape tension, torques, etc., why couldn't you offer a six-month warranty, or as you discover your repairs getting better, a one-year warranty on the VCR? (Exclude lightning, small kids, fire and problems outside of the normal wear and tear category).

Even if you see 5 or 10 percent of these machines back within the one-year time limit, the extra business will more than offset the extra work required to correct the problems in these few machines. You may be pleasantly surprised to find the return percentage to be far less than 5 percent when you eliminate guessing about mechanical wear and think about the good will you will reap.

Make Money On Abandoned VCRs

A fair percentage of customers will choose to abandon their VCR after they hear the repair estimate. This may actually be good news. You may make more money on these VCRs than if you had repaired it for the customer. By repairing the VCR, and again offering a one-year warranty, you will find that selling these machines for $95 ($75 low-end to $150 for a high-end 4-head Hi-Fi unit) to customers who bring in their broken VCRs may help both your customers and your bottom line.

If you sell a reconditioned VCR for $95 that would have brought $65 for repair, you make an additional $30 (plus you may pick up his broken unit for free, since the customer now has a working unit and may not want to pay to have his unit serviced). Being able to accurately measure remaining video head tip protrusion on these reconditioned units will help to determine how much use they've already had, and thus how much more they're likely to endure.

19 VCR Service Centers: Tips on Remaining Profitable

New heads typically protrude 40 microns. When the heads are worn out they're down in the 10 micron range. A used machine with 25 microns remaining is in a totally different category than one with only 15 microns. The warranty and price may both be based on this number.

Find More Business

You may be doing lots of things right, but if you don't have enough VCRs to service, you won't be profitable. Coupons in local TV program listings, although fairly expensive, have proven to be an excellent method to reach your target audience. Other good programs we've heard of are flyers posted at video tape rental stores.

No one is more interested in proper mechanical repairs than these store owners, since tape tensions, torques, guide heights, carriage latching problems, and similar problems, cause the vast majority of their ruined tapes. Rental stores need to have access to a reliable repair shop since they lose many tapes due to defective VCRs.

You might want to offer a Saturday "clinic," where you set up in their store and for $15 (or some nominal fee) clean the head (proper method, not just a cleaning tape), measure hold back tension, a few critical torques, reel heights, guide heights, RF envelope, and for $5 more you can even measure their remaining head wear. Even if their VCR is perfect, they'll know who to bring it to if they experience a problem.

If their VCR is in need of repair, you'll probably be able to write up a ticket to take it back to your service center. Often you can stay fairly busy just from word of mouth of these "customers," since they can now inform friends and family of an "expert" they know—you.

Another low-cost method of finding more business is "discount" coupons placed under windshield wipers at malls, baseball games, or anywhere there are a lot of automobiles. Have coupons printed 3 or 4 up on an 81/2 x 11 sheet. This reduces cost and it appears more like a discount coupon. Remember, almost everyone has at least one VCR and may know others who are having problems.

Expensive Products Will Be Serviced

Doing repairs properly, providing a worthwhile warranty, and generating new business is the only way you can remain profitable in the current low-price VCR service market. There are a growing number of more expensive four-head and Hi-Fi VCRs and camcorders that will not be viewed as disposable, and most customers who pay even $150 for a low-end machine will look favorably at a $65 repair rather than another $150. The consumer electronics service business should improve very soon with the new technologies of wireless cable, flat-screen TV, projection TV, and other high dollar consumer products that will help support the independent servicer.

VCRs will certainly be with us for many years. There is no other method that can provide six hours of video recording for $2 or $3 even dreamed of at this time. The VHS format is also being used for time-lapse recorders in convenience stores, banks, etc., and due to the high data memory, companies are also using it for digital audio recording, and computer data backup. Don't give up on VHS, but if you're going to be profitable doing service, perform the service correctly, and stand behind your work.

Chapter 20
Camcorder Servicing
By the ES&T Staff

For those who are new to camcorder servicing, and for those who have been servicing camcorders for some time, we'll start with one basic statement: a camcorder consists of both a camera and a videocassette recorder. It's easier to understand and service these products if you can get a picture of the camcorder as a whole, and appreciate just what a complex product it is. Its electronics and mechanical systems are similar to those that you'd find in a VCR, and its optics are like those that you'd find in any camera, and include the electronics circuitry and mechanical gadgets that adjust the optics for such functions as zoom, focus, and autofocus. In addition, a camcorder also contains a small video monitor, called the electronic viewfinder or EVF, that allows the user to see exactly what he's capturing on film.

Of course, there is an important subset of circuitry that a VCR contains that a camcorder does not: a tuner.

It's impossible to do justice to camcorder servicing in a relatively brief magazine article, so here's what this article will do:

- explain the functions of some of the circuits that are unique to camcorders,

- emphasize the power circuits, and describe some of the possible failure modes caused by problems in the power circuits. This article will be based on the Hitachi VM-2400 video camcorder.

The Sensor

One of the problems in dealing with new technology, such as camcorders, is the difficulty that design engineers have in coming up with terminology that is descriptive of the functions performed by a given component or circuit seg-

20 Camcorder Servicing

ment. So, engineers who work for different manufacturers, even within "Japan, Inc.," may call the same things by different names. Or, even if they call things by the same name, the technical literature from different manufacturers may place different degrees of emphasis on the terms.

For example, in the service manual for Hitachi VM-2400, the circuits and components that are used to convert the image formed by the camera lens into an electrical signal and condition it so that it can be processed by the signal processing circuitry is called the "sensor." Flipping through the manuals of other manufacturers showed that while some seemed to call it the sensor, others either did not, or simply did not employ that terminology.

Another possible source of confusion here is that camcorders have, as do VCRs, other totally unrelated sensors, which sense excessive moisture (dew sensor), the end of tape, and other conditions that warrant shutting down of the mechanisms.

In this camcorder, the "sensor" is the image sensor, which primarily consists of the CCD imager (IC001), the sensor driver (IC005), the pre-process circuits (IC003) and other support circuitry. All of this circuitry is located on the sensor PCB.

The CCD

The beginning of the entire process of recording a scene using a camcorder is the conversion of a visual (light) image to an electronic signal. In most consumer camcorders, that's accomplished by using charge-coupled device array (CCD). A CCD is a device that outputs an electrical signal when light falls on it. A typical two-dimensional camcorder CCD array may be able to generate thousands of pixels (picture elements).

The picture information signal generated by the CCD is then transmitted to the pre-process IC where the video signal is conditioned and separated into the luminance (brightness) and chrominance (color) signals.

These signals are sent to the process section where they are processed, then output to the main luminance/chroma section, then ultimately to the video head where they are recorded onto the tape.

Camcorder Servicing **20**

Troubleshooting the Camcorder

In any system as complex as a camcorder, there are literally hundreds of things that can go wrong and cause improper operation. Some of these things may cause little more than slight misadjustment of the color, or a little noise in the picture. Other things may cause the VCR to cease functioning altogether.

In this article, we will focus on the power generation and distribution circuits in the camcorder. Here are some steps to take when you run into a camcorder problem.

Camcorder Doesn't Operate At All

The power for a typical camcorder is supplied either from a charged battery or from an external power supply. There is no power supply circuitry, as such, on a camcorder. That is, the circuitry that steps down the ac line voltage rectifies and regulates it. The camcorder is dc only.

When the operator is using the camcorder in portable mode, the power comes form the battery. When the camcorder is plugged in, the power is dc from the adapter. The adapter is where the voltage transformation, rectification and regulation take place.

Power for the various functions and circuits of the camcorder takes a tortuous route. The raw B+ comes either from the battery, via the dc light control pc board, then via a 2A fuse to the system control board for distribution, or via the dc input jack on the system control board, then via the 2A fuse back to the system control board.

In addition to appearing on the system control board, this unregulated 9.6V (A9.6V) is routed to the switching regulator, where it is used as a source for the regulator, which produces several regulated voltages which are used as supplies on the system control PC board, and/or further distributed to other circuitry within the camcorder.

The switching regulator also develops the B+ for the capstan and cylinder. This signal is routed to the main servo printed circuit board.

157

20 Camcorder Servicing

From there, the cylinder and capstan B+ are route to the capstan motor drive pc board, and the cylinder B+ is routed to the cylinder motor drive circuits.

The A9.6V is also routed to several other boards, including the trouble sensor board, and the process (encoder) circuit board. From the process (encoder) circuit board, the A9.6V is, in turn, routed to the dc-to-dc pc board. Here this voltage is used as a source of supply to generate all of the voltages for the camera section: C-8V, C15V, C9V, and C5V.

When power supply problems occur in a camcorder, tracking down the problem and correcting it becomes to some extent an exercise in following circuit traces, and detective work.

Tracking Down Power Problems

Based on this analysis, it's possible to isolate power problems by performing a little detective work. For example, if the camcorder has power, and everything seems to be operating, but when some or all of the camera functions, such as zoom, or focus, don't operate, there is likely a problem with the dc-to-dc converter.

One possibility is that the A9.6V source voltage is not arriving at the dc-to-dc converter input plug. Check at pin 1 of CN1DD. If there is voltage here, one of the foils on the board may be burned or cracked, or one or more components on the board may be faulty. If there is no voltage at this point, trace the A9.6V circuit back until you find the point where the voltage appears.

If only one of the camera functions is not operating properly, the problem may be that the motor that controls that function is defective, or the voltage that controls that function may be absent. Check the input to the inoperative motor. If the input is present, check the motor. If the input is absent, track back to see where it disappears.

Suppose the opposite is true. If the camera is totally dead, the main power is either not getting to the camera at all, or is being shortstopped somewhere. If the camcorder is being operated on a battery, the obvious steps here are to check the battery, operate the camcorder with a different battery, or operate the camcorder using the ac adaptor.

If the camcorder remains totally dead, try tracing the power path. Check to see if there is A9.6V at pin 4 of the dc control board. If not, is the fuse blown? If there is power here, check step by step until you reach the inputs of the dc-to-dc converter and the switching regulator.

Divide and Conquer

A camcorder is a complex device, but its circuits are pretty much neatly divided into functions, and the circuitry for each function is grouped on a single circuit board. The power circuit, on the other hand, snakes throughout the entire system. In general, troubleshooting of camcorder power circuits is accomplished in the familiar manner of starting from a point where the power is absent and moving toward the power supply until you find the point where it is present, or starting from the power supply and moving down stream until it disappears.

Along the way, it helps if you can read the route maps and signposts that the manufacturer has put there for you, such as plug and connector numbers and circuit designations.

Chapter 21
Hitachi VCR Repairs
By Victor Meeldijk

A Hitachi Model VT-M241A that I was called on to service exhibited a wavy looking picture (*Figure 21-1*) and low sound level. The sound level would increase then fade out when the machine was put into Fast Reverse then Play mode. When I opened the unit, I found that it was relatively clean inside and seemed to be loading tape properly. Minor adjustments to the audio head stack alignment improved the audio level.

Figure 21-1. *When I first operated the Hitachi model VT-M241A to evaluate its condition, it produced this wavy picture.*

A close examination of the chassis revealed that the right tape arm was not fully seated in the stopper (*Figure 21-2*). Lubricating the tape loading arm tracks and then the gear assembly did not fix the problem, the arm still would stick just before seating. I remembered that the owner said that it had jammed once and he had had to manually extract the tape.

Checking both tape loading arm paths, I found that by pressing on the metal of the right loading arm chassis path (*Figure 21-3*) the metal moved down, this did not occur on the left side. Bending the metal into its proper position fixed the video problem. However, on system check-out the VCR hung-up, just as the owner described, and the machine went out of alignment. I traced the problem to an intermittent mode switch (pin one lost continuity). Hitachi does not

21 Hitachi VCR Repairs

sell the switch, so I ordered a complete loading block assembly (p/n 7468854). After I installed the new loading block I set about to realign the VCR.

Figure 21-2. The picture of Figure 21-1 was wavy because the right tape arm was not fully seated in the stopper (easily missed when looking at the chassis from the front of the machine).

Figure 21-3. Pressing on the metal of the right loading arm guide revealed that it was slightly raised, causing the arm to bind.

Realigning the VCR

The technical manual for this model is no longer available, so to realign the machine I used a Hitachi Model VT-M250A, which has a similar chassis, for visual reference, and a manual obtained from some helpful readers through a Reader's Exchange item in this magazine. Thanks again to Kevin Wood and Dave Garber of Pyxis Technical Services of Halifax, Nova Scotia for their help.

To keep this article short, I will not go into detail on another problem that I encountered during this service procedure: the cylinder head motor did not operate. In brief, however, by using signal tracing, I found that the clock signal from the system microprocessor (IC901) was not getting to servo IC601, pin 53. The cause of this problem was a damaged printed circuit board trace, that occurred during the handling and troubleshooting process. This can happen to you, especially with some of the thin traces in high density areas of circuit cards.

Repairing a Hitachi Model VT-M250A

At the same time as I was servicing the Hitachi Model VT-M241A, I was also working on another VCR, a Hitachi Model VT-M250A that was "eating" tape. The problem was that the tape was not being drawn into the cassette because the take-up reel was not turning. The machine would also not work in Fast Forward, Fast Reverse or Review mode.

In disassembling the drive gears I found that the fix washer had spread apart and was no longer holding the gears in place (*Figure 21-4*). After I replaced the washer the machine operated normally. However, a few weeks later the machine was back with the same problem. The new washer had spread open. To make sure that this failure did not happen again, I replaced the washer with a metal E-clip (*Figure 21-5*).

Figure 21-4. The Model VT-M250A was eating tape because the fix washer had spread apart and was no longer holding the gears in place.

Figure 21-5. After the VCR returned with the same problem a few weeks after the initial repair I replaced the fixed washer with a metal E-clip to hold the gears in place.

Chapter 22
Magnetic Recording Principles
By Lamar Ritchie

An article in the January issue described in general the principles of magnetic recording, and provided specific details on audio recording. This article provides information on recording of video, with detail on each of the popular consumer video formats: VHS, Beta and 8-mm.

Video Recording

The writing speed for video recording must be far greater than that for audio. To record the higher frequencies using the same tape velocity that is used for audio would produce far too short a recorded wavelength. For one thing, it would be impossible to manufacture tape heads with a gap small enough.

Moreover, there is a problem because of the broad frequency range needed: 30Hz to 4.5MHz, or about 18 octaves. A recording across this number of octaves would result in a recorded dynamic range of 110dB, far too much to equalize.

Frequency Translation

The first thing that must be done to record video is to translate all frequencies upward. If all video frequencies are moved upward by 4MHz, this would give a band for the video frequencies of approximately 4MHz to 8MHz. This would be only one octave.

Since the frequencies must be moved up, it was decided to record the video as an FM signal. The video is not recorded directly, but modulates an FM carrier. Using this method helps to prevent amplitude variations in the record/playback process from degrading the video.

22 Magnetic Recording Principles

Another problem is achieving the writing speed required to increase the recorded wavelength to manageable levels. Longitudinal recording, such as used for audio, is not feasible; it would consume tape too fast. For example, the writing speed for the present VHS format is approximately 6m/sec, and for Beta it is approximately 7m/sec. All video machines, therefore, provide the high writing speed by "scanning" the tape (moving the heads past the tape) at high speed. The tape moves at all only to present a new surface for this process.

Recording the Audio

On most VCRs the audio is not recorded along with the video, but uses separate audio heads and conventional audio circuits. A narrow strip of the tape is used for the audio tracks. Some newer consumer machines, however, do place the audio on an FM carrier also and place the audio heads alongside the video heads to scan the tape. Hi-Fi recordings with excellent quality can be produced this way. Our discussion, for now, will be centered on recording and playback of video signal.

The Technology

The technical sophistication required for the magnetic recording and playback of a video signal, particularly a color signal, is immense.

It is little more than a decade that color video recorders have been available for the consumer, at least at a reasonable price and performance. The actual principles have changed little with time. Mainly the quality and manufacturing improvements have created improvements in the performance of these machines.

All video tape recorders, from the most advanced professional models to the bottom-line consumer models have some things in common:
- A rotating "scanner" to move the heads past the tape to obtain a high writing speed.
- Servo control to put the heads in the exact position required as they scan the tape. To do this either the scanner motor, capstan motor, or both can be servo controlled. To obtain high enough quality for color pictures, usually both must be.

Magnetic Recording Principles **22**

- An FM modulator (record) and an FM demodulator (playback) for the video.
- A conventional audio system using space on the same tape.
- A system to record a "control track" longitudinally (like the audio) to use as a reference in positioning the scanner at the correct place at all times.

Professional Video Recording

The highest quality video recorder, used for TV broadcasting, is the Quadruplex, or "quad" video recorder. It uses two-inch video tape (reel-to-reel in most) and a scanner with four heads, as indicated in *Figure 22-1*.

The scanner rotates at 14,400RPM and uses a vacuum to form the tape into a bowed shape around it. Because 14,400RPM is 240 revolutions per second, the scanner rotates 4 times per field. This means there are 16 passes of a video head to record one field.

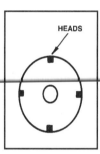

Figure 22-1. The video heads are arranged in this fashion on the quad scanner cylinder. Tape motion is perpendicular to the plane of the cylinder.

The tape speeds used are 7 1/2 inches per second and 15 inches per second, selectable. There are two versions, depending on the frequency of the FM carrier used: high band and low band. The recording format, the manner in which the various signals are placed on the tape, is shown in *Figure 22-2*.

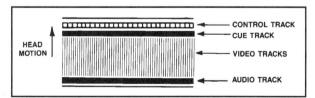

Figure 22-2. The information is recorded in this scheme in the quad videotape format.

167

22   *Magnetic Recording Principles*

Helical Scan

A simpler, less expensive way of scanning the tape is to use a "helical scan." All modern consumer VCRs use this method. Improvements have been made so that many professional recorders also use it.

In this method, the video head describes a helix as it rotates, which creates a recorded track that is a series of diagonal stripes across the tape. To accomplish this using one head, the tape would have to be wrapped entirely around the scanner. In practical VCRs, the tape is wrapped 180 degrees and two video heads on opposite sides of the cylinder are alternately switched on, to insure head to tape contact at all times. One diagonally recorded track can contain a complete field.

Two earlier methods that were used to provide a 360 degree wrap of the tape for machines using a single video head were the alpha wrap and the full omega wrap. These systems were so named because the pattern of wrapping of the tape resembled the respective letter in the Greek alphabet. The alpha wrap looked like the Greek letter α and the omega wrap looked like the Greek letter Ω.

The 360 degree wrap had an inherent problem: it was difficult to insure that the single head stayed in contact with the tape at all times. It also made threading the tape difficult. All modern VCRs use a switching system with two video heads. Tape wrap is for a little more than 180 degrees. This insures that one head, at least, is always in contact with the tape.

One of the earliest methods, used with reel-to-reel type recorders was the half-omega wrap. Modern VCRs use a threading method that is similar to the half-omega wrap. *Figure 22-3* shows roughly how the tape was wrapped, or threaded, around the video heads for these machines.

To form the helix, the tape has to cross the scanner at an angle. This can be done one of two ways. Earlier, reel-to-reel machines tilted the tape path by mounting one reel higher than the other. Usually the supply reel was mounted above the take-up reel so that the tape had to move downward as it crossed the head. This formed the helical tracks. The actual tracks on a straight piece of tape are long lines slanted at an angle, and these machines were sometimes called "slant track" machines.

Magnetic Recording Principles 22

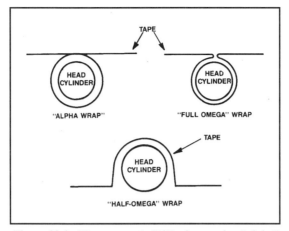

Figure 22-3. The scanner in VCRs that employ "alpha" wrap and the full "omega" wrap, each so-called because the arrangement of the tape resembles the respective Greek Letter, uses only a single video head. All modern VCRs use the "half-omega" wrap scheme, which requires two video heads spaced 180 degrees apart on the scanner.

Modern VCRs do not use this method of producing the slant tracks because it would make the cassette larger and more expensive. Instead, the head is tilted. This produces the slant video tracks on the tape as the head scans the tape moving straight across from supply to take-up reel. *Figure 22-4* illustrates these concepts and the shape of the recorded video tracks that result.

Most modern VCRs use one of the two basic formats; Beta and VHS (VHS stands for "Video Home System," or "Vertical Helical Scan," depending on who is talking). They are similar, but enough differences exist to make them incompatible with each other. *Table 22-1* shows the tape formats for Beta and VHS.

Note that the formats are similar, and both formats use 1/2 inch tape, but the threading method and hence, the cassette cases, are different. They are not similar enough, however, to allow a Beta machine to play a VHS recording even if you removed the tape and put it in the correct type of cassette.

The Beta format is actually somewhat better in picture quality than the VHS format because of the longer track length. The VHS format, however, has become much more popular because it allows for a longer recording time.

22 Magnetic Recording Principles

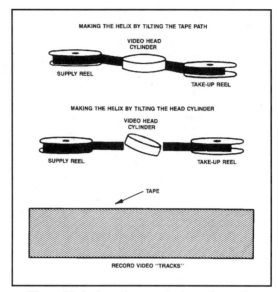

Figure 22-4. One of two methods can be used to create the helical scan pattern of the video signal on the video tape: tilt the tape path or tilt the head cylinder. Modern VCRs tilt the head cylinder because this method can use a smaller, less expensive, cassette. A general idea of the physical arrangement of the video tracks on the tape is shown at the bottom of the figure.

Table 1. Beta and VHS tape parameters

Format	Speed	Track	Track
Beta 1	1.6 in/sec(40 mm/sec)	58 microns	117 mm
Beta 2	0.8 in/sec(20 mm/sec)	29.2 microns	117 mm
Beta 3	0.53 in/sec(13.3 mm/sec)	19.5 microns	117 mm
VHS—SP	1-5/16 in/sec(33.35 mm/sec)	58 microns	97.3 mm
VHS—LP	21/32 in/sec(16.67 mm/sec)	38.5 microns	97.3 mm
VHS—EP	7/16 in/sec(11.12 mm/sec)	19.2 microns	97.3 mm

Table 22-1.

Magnetic Recording Principles   22

In both VHS and Beta machines, the scanner is most often referred to as the cylinder. It may also be referred to, in some service manuals, as the "upper cylinder," the "lower cylinder" being the motor that rotates the scanner. In older machines it was called the "drum" and is sometimes called this in newer machines.

U-Matic

Another type of cassette recorder that came into common use as a commercial video cassette recorder was the U-Matic format. The U-Matic machines are still used in industry and are popular at TV broadcast facilities. It is a format that uses larger tape (3/4-inch) and a larger cartridge. The scanner is much larger producing longer tracks, and a faster tape speed is used so better, more stable pictures are possible. This machine was the forerunner of the Beta machine and, mechanically, the Beta machine is a miniaturized version of it.

As for consumer VCRs (Beta and VHS), FM is still used to record the luminance video. In the Quad recorder, the complete video (Y and C signals) modulates the FM carrier. This is possible with the Quad machine because of its quality.

The head in a Quad machine alone may cost several thousand dollars and the complete recorder almost $100,000 dollars. It is still not feasible, however, to record the color signal along with the luminance for the consumer VCR.

Recording Chroma and Luma

To see why the color (chroma) and luminance (luma) can't be used together to modulate the FM carrier in consumer VCRs, consider these facts. The head cylinder, only two to three inches in diameter, scans the tape producing tracks that are roughly five inches long. A single track contains one complete field of video. Thus each horizontal line of video is recorded on a length of tape that is 5/262.5 of an inch, or about 0.02 inches long.

Remember now about head gap width, writing speed, and recorded wavelength.

If the signal has color on it, a 3.58MHz signal produces about 227 cycles per horizontal line, so each cycle of the color occupies about 0.02/227 inches, or about 8.8 millionths of an inch. There will be vast errors if the heads are only millionths of an inch in the wrong place at any time. A change in phase of only

22 *Magnetic Recording Principles*

a few degrees from one rotation of the head to the next is intolerable—color oscillators in TV sets could not follow it.

Overcoming the Problems

Before video can be recorded suitably, there are many problems to overcome:

- Variations in tape tension
- Tape stretch
- Dropouts (oxide bad or missing)
- Head-to-tape contact at the high writing speed
- Variations in capstan speed and rotation speed of cylinder
- Providing electrical contact to rotating heads
- Crosstalk between adjacent tracks
- Switching between the two video heads at the correct time.

The Tension Regulator

To correct for variations in tape tension, a tension regulator is used. This is a mechanical device that will be the first component in the tape path. It usually is a metal band with a felt strip on the inside that provides variable "braking" for the supply reel. If the tape has too much tension, a loop in the tape will tighten, pulling on the band to release it slightly and allow the supply reel to turn easier, lowering the tension. If there's too little tape tension, the tension regulator compensates to increase it.

The Dropout Compensator

To make dropouts less objectionable, all VCR's have a dropout compensator, or DOC. This is a circuit that delays the video for $63.5\mu sec$, and thus always has the previous horizontal line output from it. The circuit senses if the video signal disappears (there is a dropout) and, if so, switches in the video from the previous line. The eye cannot detect momentary dropouts of only one or two lines because adjacent lines are very similar.

Magnetic Recording Principles **22**

Head-to-tape Contact

As for head-to-tape contact, the heads are made to protrude very slightly into the tape, that is, they project slightly from the surface of the cylinder they are mounted on. The biggest problem here is tape contact with the cylinder itself, which must rotate at 1800 RPM. Actual contact with the cylinder would destroy tape and cylinder in a short time. Lubricants cannot be used, as they would destroy the tape.

What is done is to put fine grooves in the cylinder. Before the tape is pulled to the cylinder in the thread operation, the cylinder is started to rotate. This creates an air cushion because the grooves and the tape actually do not touch the cylinder, but ride just above its surface on the air cushion. If anything were to fill up these grooves, even a tiny amount of oil, the cylinder and tape would seize, possibly destroying the heads and the tape.

The use of an FM carrier for the luminance video also helps to minimize variations in head to tape contact because the FM demodulator is not sensitive to amplitude variations.

Getting the Signal To and From the Heads

To couple the video signal to and from the rotating heads, older machines used slip rings and brushes. These gave many problems because of dirty, bent or worn contacts. All VCRs now use rotary transformers to couple the energy to and from the heads. One circular winding is in the rotating upper cylinder, the other is in the stationary lower cylinder. The circular windings are precision wound with primary and secondary windings that keep a constant relative position to each other.

Speed Control

To minimize speed irregularities, the cylinder and capstan motors are servo controlled. Switching between the heads is accomplished at the proper time by generating a reference pulse from a pulse generator in the rotating cylinder and comparing this pulse to the vertical sync pulse.

22 Magnetic Recording Principles

Still, the heads cannot be positioned accurately enough to allow the color to be demodulated and used "as is," even in the expensive Quad machines.

The phase and frequency variations that remain, caused by the small short term changes in writing speed, are called "jitter" components of the signal. These timing irregularities are very harmful to the color signal because this signal is so phase dependent.

All video recorders that use color must have a type of "velocity compensation" whereby two signals are created with the same jitter components and heterodyned together to produce a stable difference signal. For broadcast use, a time base corrector is used to stabilize the longer term variations in the signal frequencies.

Modulation and Demodulation

VCRs are not capable of recording the color in the same band of FM frequencies in which the luminance is recorded. The FM modulation and demodulation would create beat interference because 3.58MHz is within the band of FM frequencies used. Quad machines can do it because their FM carriers are up in the range of 8MHz to 12MHz.

The cost would prohibit consumer equipment from using this method. Instead, the color is moved out of the way. It is moved down in frequency, below the FM luminance frequencies. This used to be referred to as the "color under" or "heterodyne" color system. This enables the limited bandwidth of present VCRs to record color separate from the luminance, preventing interference between the two. It does, however, limit the bandwidth of the color that is recorded. The actual frequencies that are used for the two systems are shown in *Figure 22-5*.

The two sets of frequencies given for the VHS luminance signal have to do with a method of preventing crosstalk and will be discussed later.

Another Beta version called "Super beta" centers the FM luminance at 5.17MHz for greater picture detail.

Magnetic Recording Principles 22

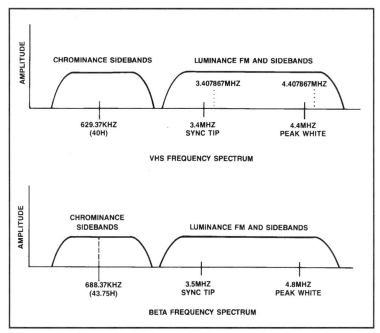

Figure 22-5. The VHS frequency spectrum is somewhat different from the frequency spectrum for Beta.

Mechanical Differences

In addition to the electrical differences, there are mechanical differences between VHS and Beta machines. To get the minimum 180-degree wrap around the head cylinder, the tape must be pulled from the cartridge and "threaded." These machines use different methods, as shown in *Figure 22-6*.

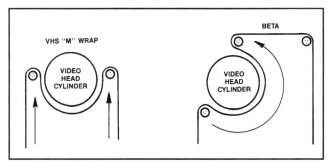

Figure 22-6. The tape wrap schemes are different in the VHS and in the Beta VCRs.

22 Magnetic Recording Principles

Other differences, in comparing Beta to VHS are:
• Beta uses a smaller cassette. This is an advantage in that it is more compact, but it holds a shorter spool of tape so there is less record time.
• The thread operation occurs as soon as the cassette is inserted in the Beta machine, and stays threaded at all times when the cassette is in the machine. This gives better control over tape movement and allows faster operation between modes, such as from stop mode to play mode. It is, however, a little more damaging to the tape during FF and REW operations. (A few "instant play" VHS machines are now doing this also.)
• Beta video tracks are a little longer, as mentioned earlier. This allows for a little wider FM bandwidth for the luminance, and therefore resolution that is a little better.

Improved Formats

Recently, new formats for both VHS and Beta have been developed that provide better picture detail (better resolution). As a comparison, broadcast TV signals have a resolution of about 330 lines. VHS VCRs use a chrominance signal at approximately 629KHz that trails off at about 1.13MHz. The luminance extends from there, up to about 4MHz, giving a bandwidth of about 2.9MHz. This produces about 250 lines of resolution.

Beta resolution is slightly better, at about 260 lines of resolution. The newer Super beta has a bandwidth of about 3.2MHz. that gives a resolution of just under 300 lines.

As mentioned, recent advances have created new formats that are superior in resolution, but unfortunately incompatible with the old formats. The two new formats are Super VHS and Extended Definition BETA (ED BETA).

Actually, there are three basic formats, but the third, 8-mm, is not yet in common use for VCRs, although it may be in the near future.

The newer Super VHS has a color signal identical to the signal used in the old system. The difference is in the FM luminance. It uses a higher band of frequencies, allowing a wider bandwidth and thus better picture detail.

The sync tip and peak white frequencies for Super VHS are 5.4MHz and 7MHz. This produces a resolution capability that exceeds 400 lines. This is much better resolution than a broadcast TV signal, and better than most TV receivers are capable of displaying. To see the picture with full quality requires the use of a color monitor with a special S-VHS input jack. The S-VHS connector has separate cables for the luminance video, color signal, and the audio channels. This provides much better quality because the Y and C signals are separate and the high frequency luminance cannot beat with the C signal and produce beat interference. These monitors are capable of displaying typically 400 to 500 lines of resolution.

Compatibility Considerations

To provide compatibility, S-VHS VCR's have built in record and playback circuits for standard VHS. For recording, a switch can select the desired format. For playback, the circuits automatically detect whether the signal is VHS or S-VHS. The use of S-VHS requires special higher quality tape that is capable of recording the S-VHS signal. The signal-to-noise ratio for the luminance in standard VHS is a little better than 40dB, and can be a few dB higher for S-VHS. The color signal is, however, improved only slightly. It is recorded exactly the same and what improvement is noted is because of the higher quality tape.

The newer ED Beta provides much more resolution than even S-VHS; over 500 lines. As with S-VHS, the color signal is recorded the same and the luminance FM band is much higher. The sync tip and peak white frequencies for ED Beta are 6.8MHz and 8.6MHz. Moreover, ED Beta uses a special tape, that is even more demanding than S-VHS, that uses extremely fine metal particles instead of oxide particles.

The Favorite

At present, the clear favorite standard is VHS. The main reason that VHS became the favorite was the decision to use a little larger cassette and thus more tape, to provide a longer play/record time. It is difficult to predict what the future will be, but S-VHS will probably become the most popular standard in the next few years. Although S-VHS is not the superior format for picture quality, it still provides the longest play/record time and S-VHS VCR's are being built that will play standard VHS recordings. This insures that the consumer's tape library will still be usable with the purchase of the S-VHS video cassette recorder.

22   *Magnetic Recording Principles*

The 8-mm Format

As mentioned earlier, there is a third basic tape format, but it is not as popular and is used primarily (at present) for camcorders. This is the 8-mm format. The original 8-mm format provides better picture quality than either VHS or Beta, including the color signal. The 8-mm format was developed to provide a much smaller tape cassette for use in camcorders. 8-mm tape is about one-third of an inch wide, only slightly larger than reel-to-reel and 8-track audio tape, but of a much higher quality, using metal particles like the ED Beta tape.

The 8-mm cassette is not much larger than the standard audio cassette. The format uses a color signal at 743KHz, and a luminance signal with sync tip and white peak frequencies of 4.2MHz and 5.4MHz. Picture resolution is between that of VHS and S-VHS.

Recently, a new 8-mm format called Hi-Band 8-mm, or "Hi–8," was developed. The color is the same and improvement was provided by moving the luminance FM frequencies up, as it was for the others. For Hi-Band 8-mm, the sync tip and peak white frequencies are 5.7MHz and 7.7MHz and provides roughly the same resolution as ED Beta.

8-mm does not use an audio track. Instead, all 8-mm machines use the hi-fi sound system in which the sound modulates an FM carrier and is recorded by heads mounted on the rotating head cylinder. This FM carrier is located in the gap between the chroma frequencies and the luminance FM frequencies. This applies to the hi-fi machines in the other formats as well. This provides for sound that has as wide a frequency response and almost as good a dynamic range as CD players.

It can be argued that the 8-mm format is the best of all, but its popularity at present is limited. It could be that the choice of the name is part of its problem. There was an 8-mm film format that was the standard for home movies before the advent of VCRs and camcorders, and some of the public may be confusing the two, believing the 8-mm camcorder to be a film camera. Of course, the lack of prerecorded tapes in the 8-mm tape format is a definite handicap.

Chapter 23
Magnetic Recording Principles: Video
By *Lamar Ritchie*

Articles in the January and February issues described in general the principles of magnetic recording, and provided specific details on audio recording as well as recording of video, with some detail on each of the popular consumer video formats: VHS, Beta and 8mm.

This article will provide further detail on video recording. We will briefly discuss the circuits and methods for VHS and Beta only, but the 8mm will be similar.

Azimuth Recording

For the standard audio, to prevent crosstalk between channels, guard bands on the tape are used. Earlier machines used guard bands between video tracks to prevent crosstalk between them. In order to reduce tape consumption, VHS and Beta machines do not use guard bands. In fact, at slower tape speeds, the video tracks may actually overlap somewhat.

One technique that is used to minimize crosstalk at the higher luminance FM frequencies is to use azimuth recording techniques. Azimuth means the vertical angle of the head gap. The two heads are given opposite azimuth. This video is illustrated in *Figure 23-1*.

For VHS, as shown, the head azimuths are + and -6 degrees. For Beta, these azimuth angles are + and -7 degrees.

As the heads are alternately switched in they will create video tracks in which alternate tracks have the opposite azimuth. An adjacent track then will have the wrong azimuth for the video head. It will "straddle" the magnetic poles on this track, causing it to cross north and south regions at the same time, producing cancellation for this signal.

23 Magnetic Recording Principles: Video

This method works for the higher frequency FM signals but not as well for the lower frequencies at which the color is recorded. The longer wavelengths produce longer magnetic poles on the tape, and the gap can still be within a single pole. *Figure 23-2* illustrates this concept.

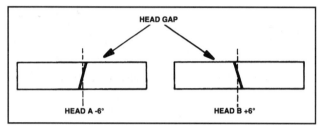

Figure 23-1. Azimuth recording techniques, the placing of the two video heads at slightly diferent angles to the tape path, minimizes crosstalk at the higher luminance FM frequencies.

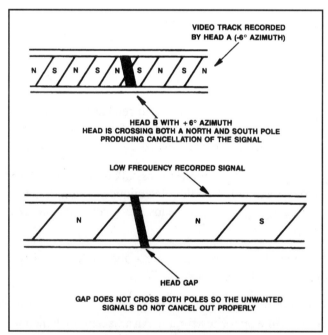

Figure 23-2. Longer recorded wavelengths produce longer magnetic poles on the tape. In this case, the magnetic head gap can still be within a single recorded magnetic pole, causing crosstalk.

For standard play speed in VHS machines, the tracks have a small amount of separation, so this does not produce much of a problem. It is not as much of a problem with Beta machines because the frequencies are a little higher and tape speeds are a little faster. There is more of a problem at the LP and SLP (EP) speeds on a VHS machine. The tracks can actually overlap at the slower speeds.

Beat Frequencies

There are actually two problems at the lower frequencies. In addition to the low frequency crosstalk for the FM luminance, there will be "beat" frequencies developed between the main signal picked up by a head and the crosstalk frequencies from an adjacent track. The beat frequencies have no true azimuth and will appear in both channels.

One way to minimize the beat is to "line up" the tracks so that all horizontal sync pulses and video line information is at the same place on each track. Since the information is very nearly the same for successive fields, there would be little beat difference between them.

The original SP speed was engineered to be the correct speed at which the tracks lined up perfectly. Perfect alignment also occurs at 1/3 this speed (SLP/EP), but not at 1/2 this speed (LP). At the LP speed, the video sync tips are in the middle of the horizontal scanning line for adjacent tracks and greatly worsen the beat problem.

FM Interleaving

As a further method to eliminate the beat problem, FM interleaving is used. As you may know from the study of color TV principles, interleaving was used for the color subcarrier to reduce beat interference. The monochrome video information occurs mostly in clusters of energy at multiples of the horizontal scan rate. To accomplish this interleaving for the luminance FM, the frequencies for one channel (for one of the heads) is raised by 1/2H, or 7,867Hz.

23 *Magnetic Recording Principles: Video*

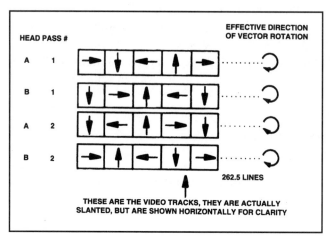

Figure 23-3. *At the lower chroma frequencies a vector rotation scheme is used to eliminate crosstalk.*

At the lower chroma frequencies, azimuth recording does not help to reduce crosstalk. Instead, a vector rotation scheme is used. What is done, in effect, is to advance the phase of channel 1 by 90 degrees each horizontal line, while delaying the phase of channel 2 by 90 degrees each horizontal line. Circuits within the color section of the VCR accomplish this (*Figure 23-3*).

Each of the boxes with an arrow represents one horizontal line. The arrows indicate the phase of the chroma signal for each line. During playback, the circuits reverse the phase angles, effectively rotating the vectors the opposite way to restore the proper phase.

To understand how this works, look at pass number 1 of head A. The crosstalk component that is picked up is that recorded by head B. In playback, as the vectors are rotated the opposite way, the information for each horizontal line will be brought back into phase. However, any crosstalk is being rotated the wrong way and will end up shifted 180 degrees each horizontal line, producing cancellation.

To produce the cancellation, the signals from each horizontal line are added to the signals from the previous horizontal line by using a 1H delay line (delays by 63.5μsec) as shown by *Figure 23-4*.

Since the main signals are in phase for each line, they add and the output is doubled. The crosstalk components, however, are 180 degrees out of phase, so these components cancel.

Magnetic Recording Principles: Video

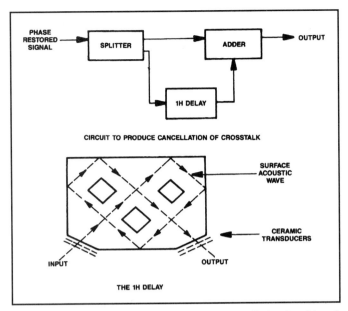

Figure 23-4. In the vector rotation scheme, cancellation is achieved by adding the signals from each horizontal line to the signals of the previous horizontal line by using a 1H delay line.

The Delay Device

The 1H delay device is normally a surface acoustic wave (SAW) device. At the speed of electrical conduction (nearly the speed of light) it would be difficult to get a 1H delay. The SAW device uses transducers to change the electrical signal to a surface acoustic wave, as shown in the Figure. The 1H acoustic delay is also used to separate the color from the luminance for the record circuits. Recall that the color is interleaved with the luminance. If the output is connected with the polarity to cause the color signals to be in phase, then the luminance will be out of phase, and cancel. The reverse is also true.

This circuit is called a "comb filter" because its output frequency response will look like a comb (see *Figure 23-5*) as the frequencies are alternately in phase and out of phase.

23 Magnetic Recording Principles: Video

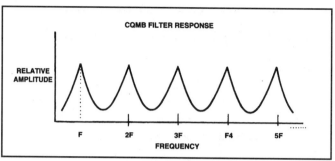

Figure 23-5. The 1H acoustic delay provided by the SAW filter is also used to separate the color from the luminance for the record circuits, using a device called a comb filter.

As mentioned before, the color signal will have "jitter" components. The color playback circuits must restore the color to 3.58MHz and remove the jitter components. A generic block of these circuits is as shown in *Figure 23-6*.

The frequencies in the figure are the "rounded" values that are normally used. The actual frequencies are:
4.27MHz (4.2679MHz)
4.2MHz (4.2089MHz)
688KHz, the Beta color frequency, is actually 688.37Kh 629KHz, The VHS color frequency, is actually 629.37KHz.
The 629KHz frequency in a VHS machine is sometimes referred to as the 40FH signal, because it is equal to 40 times the horizontal scan rate:
40 x 15,734Hz = 629.24KHz
A frequency used to rotate the vectors by 90 degrees is the 160FH signal. This frequency is used in a circuit called the 4-phase logic circuit.
160 x 15,734Hz = 2.52MHz (2.517482MHz)

This is four times the converted color frequency and when divided by 4 in the color sequencer circuits will equal the color frequency.

The Playback Signals

During playback, two signals are produced by the heads, the FM luminance and the color sidebands. These signals are very low amplitude: only about 10mV or so. The signals are coupled to the rotary transformer winding mounted in the bottom of the rotor via wires from the head cylinder. From there, they couple into the rotating winding of the rotary transformer. Induction couples these signals into the stationary winding of the transformer mounted on the lower cylinder. The signals are then coupled into a head preamp.

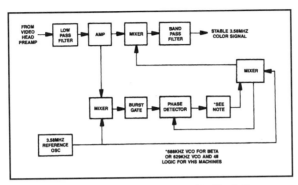

Figure 23-6. The circuits shown in this block diagram restore the color to 3.58MHz and remove the jitter components from the playback signal.

The Head Preamp

The head preamp may be physically mounted to the head's base, or lower cylinder structure. This allows for the shortest connections possible and less chance of picking up stray magnetic fields, causing noise in the signals. The preamp may also be located on the main PCB, with shielded coax cable connections from the heads. In either event, the head preamp will be well shielded to prevent noise pickup.

The main component of the head preamp will usually be an IC, containing the preamplifier, filters to separate the luminance FM and chroma signals, and the head switching circuits.

The inputs to the head preamp will be the signals from the two heads, the dc power supply voltage, typically 5V, and a 30Hz square wave (the head switching pulse) from the servo circuits. The outputs from the head preamp will be the chroma signal and FM luminance signal. Using bandpass filters, the chroma signal will be routed to the color circuits, just described, and the FM luminance will be routed to the luminance circuits. Typically, these circuits are known collectively as the Y/C circuits, or Y/C section of the VCR.

The luminance circuits contain the FM demodulator for playback of the signals, and the FM modulator for recording of the signals. These are typically

23 Magnetic Recording Principles: Video

within a single IC in modern VCRs, with built in switching circuits to select record or playback, using a digital control signal from the syscon (system controller).

The video output from the FM demodulator is fed to both the video output jacks, and the video input of the RF modulator. The RF modulator contains the circuits to generate and modulate the carriers to enable the playback signals to connect to a standard television, on either channel 3 or 4 (switch selectable).

Of course, the chroma and luminance signals and frequencies just described cannot be obtained unless the servo circuits are operating properly. These are the circuits that control both the speed and positioning of the scanner rotation and tape movement.

The Servo Circuits

The servo circuits for a VCR are responsible for a number of functions:

• Precise positioning of the video heads as they rotate. This is known as the cylinder or drum servo.
• Precise control of tape speed, synchronized with the rotation of the head cylinder, to insure that the head is allowed to precisely follow the video tracks. This is called the capstan servo.
• Generation of the 30Hz head-switching pulse.
• Fine adjustment, to allow for slightly different track length, width, and positioning for recordings made on different machines: the "tracking control."

A simplified block diagram of the servo circuits is shown in *Figure 23-7*.

To provide such precise control, the servo circuits must know the position of both the capstan motor and the cylinder motor at all times during their rotation. This is accomplished by using reference generators in each. The reference generators are pickup coils that have voltages induced into them by magnets or loops mounted within the structures. These coils generate pulses that correspond to position as they rotate. The cylinder must have two pulse generators. One generator generates a 30Hz pulse corresponding to the time when one head is leaving the tape and the other is entering the tape. This is called the head PG (pulse generator) pulse.

Magnetic Recording Principles: Video 23

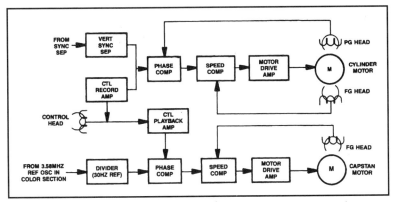

Figure 23-7. A simplified block diagram of the VCR servo circuits.

However, for precise control of positioning during the video tracks, this frequency is not high enough. A second generator generates an FG (frequency generator) pulse for this purpose. The exact frequency being used for phase control here, may vary from one machine to the next, but will be a multiple of 60Hz.

The Capstan Reference Generator

The capstan reference generator produces a capstan FG pulse. It is well to point out here that the FG pulses are not really pulses, they are sine waves. The servo circuits will precisely control the frequency and phase of these sine waves by controlling the currents delivered to the motors.

The speed of the capstan motor will be controlled by comparing the 30Hz control track pulses with a 30Hz signal derived by dividing the 3.58MHz signal from the color section. The control track pulses are obtained from the control track head which is positioned on the same physical mount as the audio head.

During record, the control track pulses were derived from the vertical sync and recorded along the edge of the tape. The capstan rotation being much slower than the cylinder rotation, the 30Hz control track pulses are sufficient for phase control (precise positioning) of the capstan movement also.

The vertical sync is the reference used for speed and phase control of the head cylinder. The sync is multiplied in frequency and fed to the phase and frequency comparison circuits of the cylinder servo section.

187

Motor Drive

As an interface between the servo circuits and the motors themselves, motor drive amplifiers (MDAs) are used to provide sufficient current to operate the motors. The motor drive amplifiers are higher power hybrid IC's (transistors on older types) that receive a control voltage from the servo. The amplifiers drive the motors to a speed that is proportional to the servo control voltage. For the capstan motor, a dc motor can be used, with the MDA supplying a dc current, the amplitude of which is controlled by the servo.

For the cylinder motor, a dc motor is not sufficient to allow the precise control needed. A multi-pole ac motor is used. Typically this motor will have three pairs of poles and use three-phase ac. The servo circuits or the drive circuits for the cylinder will vary the frequency of the voltages applied to this motor. Three voltages, 120 degrees out of phase, are developed by the circuits. This true rotating field allows much more precise positioning during the cylinder's rotation.

Precise Tracking

For more accurate tracking, most modern VCRs also use multi-pole motors and ac servos for the capstan.

To allow for precise tracking, the control head must be in precisely the right position along the tape path. To allow for correct "centering" of the head. A mechanical adjustment is provided to move it laterally a small amount. This is usually done with a conical shaped threaded nut that, as it is threaded up or down, displaces the head one way or the other. Electrical adjustments are sometimes provided on the main PCB to center the tracking. On some units there is an adjustment for each tape speed.

The System Controller

To precisely control all the functions of a VCR, a microprocessor is used. This is a dedicated microprocessor called a system controller. This circuitry is sometimes referred to as the SYSCON. Controlling the necessary functions of a VCR is quite complicated, and the controller has many functions. Refer to the simplified block shown in *Figure 23-8*.

Magnetic Recording Principles: Video 23

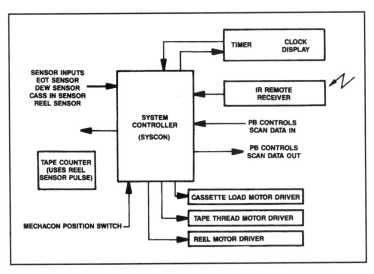

Figure 23-8. A microprocessor, called the system controller, or SYSCON, is used to precisely control all the functions of a VCR. Controlling the necessary functions of a VCR is quite complicated, and the controller has many functions.

Of course, one function of the system controller is to invoke the functions selected by the user (play, record, etc.). To accomplish this the control panel switches are usually arranged in a matrix having rows or columns that are scanned by the controller. Matrixing allows for less PCB wiring and fewer connections. The controller will thus have "scan data" outputs and inputs.

A single encoded data line is usually provided from an IR receiver to allow remote control of the VCR. The transmitter generates different data codes depending on which key function is pressed. The syscon decodes the transmitted data.

The controller has outputs to control the motors that load the cassette tape, thread the tape or turn the reel drive motor forward or reverse. Outputs tell the servo section when to start rotation of the cylinder and capstan.

Many machines were made with separate capstan motors, reel drive motors, cassette loading motors and tape threading motors. Some VCRs have mechanical assemblies that allow some motors to have more than one function. In many newer machines a single motor, the capstan motor, performs all of these functions. The mechanical arrangements that allow multiple functions from one motor are quite complex.

Switches and Sensors

In order for the controller to "know" what function is being performed and when the motor has completed performing that function, mechanical switches can be used. One such multi-position switch may be referred to as the "mechacon" switch. Alignment of this switch position is sometimes referred to as control timing adjustment.

Switches may be provided to indicate the position of the cassette tape as it loads as to when a tape has been inserted, when it is fully loaded or fully ejected. The mechacon also provides for some protection of the heads, circuits and tape.

The Dew Sensor

One type of protection that is provided is to insure that the tape cannot contact the rotating cylinder when dew has formed on the cylinder. The moisture would fill in the air grooves on the cylinder causing the cylinder to "grab" the tape. To determine when this condition exists, a "dew sensor" is used. The dew sensor is a small element whose conduction increases when its layers of wiring become moist. When the dew sensor signals to the controller that there is moisture present, the controller causes all tape motion functions to be inoperative.

The Reel Sensor

If, for any reason, the take up reel could not take up the slack as the capstan's pinch roller pulls the tape from the supply reel, the tape would spill out inside of the machine. This would ruin the tape and possibly cause other serious damage. A protection circuit is therefore provided for this condition by means of a "reel sensor."

In present day VCRs, this sensing is provided by putting mirrored stripes on the bottom surface of the take up reel. An infrared LED shines upward toward the reel. Positioned alongside it, also pointing toward the reel, is a photo detector. As the reel turns, the light will alternately be absorbed and reflected back to

Magnetic Recording Principles: Video **23**

the detector, producing square output pulses if the reel is turning. The controller senses whether these pulses are present or not. If no pulses occur within 1 to 3 seconds, STOP mode is initiated by the controller. The reel sensor itself usually has both LED and detector mounted in the same package, being a single component with four lead connections.

The reel sensor has a second function: to supply pulses as the reel turns for the tape counter.

End of Tape Sensor

Sensors are provided to tell the controller when the tape has reached the end of its travel. If the end of the reel did not signal the system to stop, the tape could be stretched or broken when it reached the end. A second purpose of the EOT (end of tape) sensors is to trigger the automatic rewind function on all VCRs. Some EOT sensors also have an auto-repeat function. A third function is to provide sensing as to whether the cassette has dropped properly into position after the cassette load operation.

For use by the EOT sensors, all tapes have a clear leader strip attached to each end. An infrared LED (in older machines, an incandescent lamp) is positioned in the center of the tape transport such that it will protrude through a hole in the center of the cassette. There is a small hole in the cassette, just inside the tape cover, through which the infrared light from the LED can shine. On each side of the transport, at the same level as the hole in the cassette, is a photodetector (normally a phototransistor).

The left detector is called the supply sensor, the right one the take-up sensor. If the end of the tape is reached, the light will shine through the clear strip and strike the sensor. This output voltage change will be sensed by the controller. When the tape is inserted, if the level of at least one of the sensors does not change, the controller "knows" that the tape cassette has not dropped into position.

23 *Magnetic Recording Principles: Video*

Tuner and Demodulator

All consumer VCRs contain a tuner and RF demodulator to provide recording capabilities for off the air and cable signals. The tuner is usually of a type identical to that found within a color TV. An RF demodulator is a circuit that amplifies the video IF frequencies supplied by the tuner and converts them to baseband video and audio. It contains video IF amps, video detector and FM sound detector. These signals connect to either a line/tuner switch of some kind, or through contacts of the video and audio input jacks.

The switch or contacts provide selection of the input signals to be recorded. Most jacks are connected such that if external video or audio sources are plugged in, the connections are broken from the RF demod providing automatic selection. The tuner and demodulator may be integrated into a single unit in some VCRs.

RF Switching

Switching must be provided at the RF output to select whether the modulator in the VCR, or the TV signal leads are is connected to the output. This switching is done electronically in modern VCR's using a control voltage and switching diodes. Some VCRs have a separate RF switch, a small shielded unit with the RF in and RF out external connections, an internal connection from the modulator, and an output to the VCR's tuner. This switch has a switching voltage connection for the selection. A VCR/TV switch on the front panel provides control, through the system controller. In addition, in most units the controller automatically switches to the VCR position when PLAY is selected.

In many VCRs the modulator and RF switching are integral. For the integral modulator/RF switch, there are the two external RF in and out connections, audio and video inputs, an output to the tuner (shielded RF cable), power supply voltage (B+) input and switching input. Sometimes the B+ voltage and switching voltage are the same voltage, this voltage energizing the modulator and causing the VCR to switch from the RF input to the modulator input.

Performance Options

Of course there are many options available for VCR's, and many circuit variations. One type of circuit being used on some machines is called an HQ (high quality) circuit. This is simply a way of sharpening up the edges of objects in the playback, giving more apparent detail. The HQ circuitry uses a "crispened comb filter" to emphasize the high frequency components of the video.

Many higher quality machines use more than two video heads. The four head machine uses only one set of two heads at a time. Another set is provided because of a compromise that must be made for having three operating speeds. The wider the video tracks, the more induced signal the heads will have. The tracks could be wider at faster tape speeds, but at slower speeds this might cause too much overlapping of tracks. For a machine to acceptably play all three speeds, the heads must be no wider than allowed for the SLP speed. This means that the playback at SP is not as noise free as it could be. A four head machine switches in another pair of heads for SP that are wider, thereby giving better performance at this speed.

Four Heads Also Eliminate Noise

Another benefit is derived from the extra set of heads. When the picture is paused, the tape has no forward motion to add to the head motion, so the heads move past the tape at a steeper angle. This causes the heads to cross over into the band between the tracks. This doesn't matter for EP because the tracks overlap, but for SP there is a space between tracks. The heads will pass into this space and video will be lost, causing a noise band in the still picture.

The SP heads on a four head machine are wider than the gap between tracks, so these heads always pick up information, and there is no noise band. The still picture, then, is noise free. This effect applies also to tape motion during cue and review operations.

For example, when the user is cuing the tape (FF during PLAY), the tape motion causes the heads to scan at a shallower angle. This angle changes to such a degree that the heads will cross through the empty bands between tracks several times in one field. Once again, this is not a problem in the EP mode, but several noise bands will appear in SP mode on a two-head machine. This will not occur, for the same reason discussed before, on the four head machine.

The Three-head VCR

A less expensive way of providing noise-free pause and cue is used on three-head machines. An extra head is placed a small distance away from one of the main channel heads, such that if the primary head misses the track, the extra head will be in the track. Switching circuits select this head at the correct time to "fill in the holes" in the signal. This head provides only noise-free special effects, and is not used during the normal play and record process. There is, then, not the overall improvement in performance at SP that there is in the case of a four head machine.

Some machines have an extra head on the head cylinder called a "flying erase head." The normal erase head erases the entire tape across its width as the tape crosses it perpendicularly. This erases the video tracks slowly, across their length. New video recorded over old will then have a noise band that runs from bottom to top of the picture until the full length of the tracks crosses the erase head. The flying erase head makes possible smooth, noise free edits of the tape by erasing the track along its length immediately preceding the recording of the track.

Many Variations

To be sure, there are many variations in circuits used in VCRs. To become proficient in repairing and servicing VCRs, it is important that the technician:

• Have a good understanding of basic electronics theory and principles. This includes knowledge of basic television theory, the composite video signal and its constituents.
• Have knowledge and experience of general troubleshooting techniques; how to check components and use test equipment such as the meter and oscilloscope to measure voltages and trace signals.
• Have a general knowledge of how the machine works. It is hoped that this material would help to accomplish this.
• Obtain "hands on" experience.
Some of the best sources for technical information about VCRs are the manufacturer's technical manuals. These can be obtained through electronics parts distributors and normally cost from $20.00 to $30.00 each. Most technical manuals contain a detailed description of the circuits and their operation in addition to complete schematics, alignment and troubleshooting procedures.

Chapter 24
Magnetic Recording Principles: Audio and Video
By Lamar Ritchie

Audio and video tape recording have provided immense opportunities for entertainment in the home. They have also provided innumerable opportunities and problems for service technicians.

An understanding of magnetic recording tape construction and recording principles can help a service technician in diagnosing a problem found in audio or video recorders.

The Principles of Magnetic Recording

The basic principle of all magnetic recording is the same, whether the information recorded is video or audio. A thin plastic "tape," coated with very fine magnetic particles such as iron oxide or chromium dioxide, is moved past an electromagnetic "head" at a constant velocity.

During recording, a variable current is sent to the head, producing a variable intensity magnetic field which in turn produces regions of varying degrees of alignment of the magnetic domains on the surface of the tape. During playback, movement of the recorded tape across the head gap causes a varying ac voltage to be induced in the head's coil.

Audio Recording Requires a Bias Signal

The audio ac voltage cannot simply be applied to the magnetic head as is. There are two reasons for this:

24 Magnetic Recording Principles: Audio and Video

- the tape's surface, being a ferromagnetic material does not have linear characteristics, and
- the head produces an output whose amplitude is not linear with frequency.

The biggest problem caused by the nonlinear characteristic of the tape's ferromagnetic material occurs at the lower end of the characteristic curve, where alignment of the magnetic domains does not start to occur until some definite, non-zero amount of magnetizing field is applied (*Figure 24-1*). To place the audio signal in the linear region, an ultrasonic bias, in the range of 60KHz to 100KHZ, is used (*Figure 24-2*).

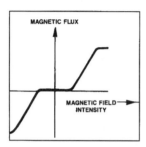

Figure 24-1. The biggest problem caused by the nonlinear characteristic of the tape's ferromagnetic material occurs at the lower end of the characteristic curve, shown here, where alignment of the magnetic domains does not start to occur until some definite, non-zero amount of magnetizing field is applied.

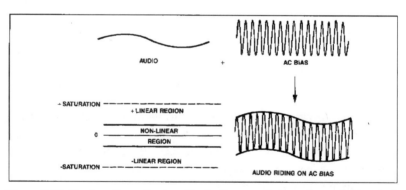

Figure 24-2. To place the audio signal in the linear region, an ultrasonic bias, in the range of 60KHz to 100KHz, is used.

The audio signal is superimposed on the high-frequency ac bias, producing a variation of the ac signal, the peaks of which do not extend into the non-linear region of the tape's characteristics. The actual signal that is applied to the head has much less variation in it than shown in the diagram. For purposes of the diagram, the variation was magnified for clarity. The actual level of the bias ac may be several volts and the audio in millivolts. The ac bias actually drives the head into magnetic saturation.

Magnetic Recording Principles: Audio and Video   24

In fact, the bias is not recorded, or recorded very little. The bias drives a particular magnetic domain of the tape around the hysteresis loop several times and as the tape moves on and the magnetic field applied to it falls to zero, it comes to rest at a certain magnetization depending on the signal current.

The Magnetic Recording Head

Figure 24-3 shows how the head places the varying alignment of the magnetic domains on the tape. The head contains a small non-magnetic gap that is placed in contact with the tape as it goes by. The two ends of the gap act as the poles of the electromagnet. The tape has a relatively high magnetic permeability and acts as a low reluctance path for the lines of force across the gap. As the lines of force vary in intensity, the degree of alignment of the tape magnetic domains is varied.

PROFESSIONAL GRADE EQUIPMENT			
REEL TO REEL RECORDERS			
TAPE WIDTH	TRACKS	SPEED (in./sec.)	SPEED (cm/sec.)
2" (5.08cm)	24-48	15,30	38.1/76.2
1" (2.54cm)	2-16	15,30	38.1/76.2
1/2" (1.27cm)	2/4/8	15,30	38.1/76.2
1/4" (.64cm)	2	7-1/2,15	19.1/38.1

CONSUMER GRADE EQUIPMENT			
CASSETTE TAPE			
TAPE WIDTH	TRACKS	SPEED (in./sec.)	SPEED (cm/sec.)
1/4"	1/2/4	1-7/8, 3-3/4	4.8/9.5
		7-1/2,15	19.1/38.1
8-TRACK CARTRIDGE			
TAPE WIDTH	TRACKS	SPEED (in./sec.)	SPEED (cm/sec.)
1/4" (.64 cm)	8	3-3/4	9.5
CO-PLANAR HUB CASSETTE			
TAPE WIDTH	TRACKS	SPEED (in./sec.)	SPEED (cm/sec.)
.15" (.38 MM)	2/4	1-7/8	4.8 (3-3/4 on some 4-track recorders)

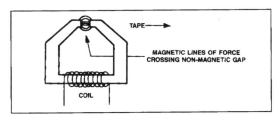

Figure 24-3. The head places the varying alignment of the magnetic domains on the tape.

24 Magnetic Recording Principles: Audio and Video

The actual size of these small magnetized areas of the tape is called the "recorded wavelength." The recorded wavelength is determined by the *tape speed and signal frequency (Figure 24-4)*.

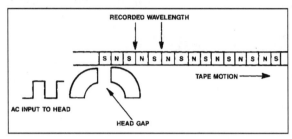

Figure 24-4. The actual size of the small magnetized areas of the recorded tape is called the "recorded wavelength." The recorded wavelength is determined by the tape speed and signal frequency.

The Head Nonlinearity Problem

One of the basics that most electronics courses teach is that the induced voltage in a conductor by generator action is proportional to the change in magnetic field intensity in the vicinity of the conductor. This is, in turn, proportional to the speed of motion of the conductor through a magnetic field. This principle also applies to the head during playback of the signal.

Higher frequency signals cause a faster change in the field for the same tape speed, and therefore, a higher output voltage from the head. As the frequency increases, the output of the head will increase by 6dB per octave. The output, then, must be equalized by a filter having opposite characteristics.

As for the range of frequencies that can be recorded on magnetic tape, the upper limit is determined by the speed of the tape motion relative to the head (writing speed), and the width of the head gap. Maximum output will occur when the recorded wavelength is twice the width of the head gap. From there on, the output will decrease and reach minimum when the recorded wavelength is equal to the width of the head gap. At this frequency, both a north and a south pole are positioned within the gap and the fields will thus cancel. *Figure 24-5* illustrates the change in relative output of the head as the frequency changes.

Magnetic Recording Principles: Audio and Video 24

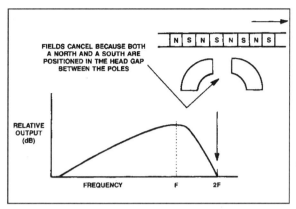

Figure 24-5. The illustration shows the change in relative output of the head as the frequency changes. The recorded wavelength is equal to the writing speed divided by the signal frequency.

The recorded wavelength of a signal is equal to the writing speed divided by the signal frequency.

Limitations in the Recording Process

Figure 24-5 indicates the response under ideal conditions and other factors have influence. Since the slope rises 6dB/octave it is important to have very little noise to extend the usable frequencies. The tape itself is the biggest factor here.

Random nonuniformities in the oxide coating cause random noise. On poor tape, oxide not bound well will come off and worsen the problem. Other factors that limit performance are:

• *"Fringing" of the field.* A smearing of the recorded pattern because the field extends a little outside of the tape.

• *Self erasure of higher frequencies.* As the domains swing around quickly, inertia carries them past the correct alignment and they come to rest at a more random point (ac fields are used to erase tapes for this reason).

• *Separation losses resulting from imperfect contact between the tape and the head.* This loss amounts to about 55dB for a tape-to-tape separation of one recorded wavelength.

24 *Magnetic Recording Principles: Audio and Video*

• *Printthrough can be a problem with magnetic tape.* In print through, the magnetic pattern on a length of tape bleeds through to adjacent wraps of the tape.

At present, the dynamic range of equalization is about 60dB to 70dB. This gives a frequency range of about 10 octaves. For audio frequencies, 20Hz to 20KHz, then, direct recording of frequencies can be used, since that represents 10 octaves.

Some of the standards for audio tape are shown in Table 1.

Multiple Tracks

To obtain maximum recording time from a given length of magnetic tape, most consumer audio recorders use multitrack recordings. Of course, for stereo sound, two channels (tracks) are required for each recording. The tracks for each recording usually take up only half of the width of the tape so that the tape can be turned over to record on the "back side."

To help prevent crosstalk between channels, the tracks are separated for some formats. The track formats for reel-to-reel and the now obsolete 8-track recorders are as shown in *Figure 24-6*.

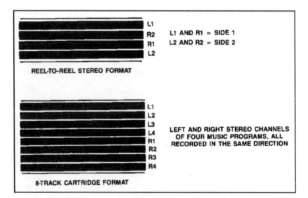

Figure 24-6. The track formats for R-R and the now obsolete 8-Track recorders.

For 1/2 track (monophonic) reel-to-reel, the top half of the tape width is simply side 1 (track 1) and the bottom half of the tape is side 2.

Magnetic Recording Principles: Audio and Video

The mono reel-to-reel is not compatible with the stereo format, because front and rear channels will overlap. The separation between tracks is not needed much, today, because of better heads and tape. The cassette format, therefore, has the tracks for each channel adjacent to each other. This arrangement provides compatibility, meaning a mono player can play a stereo recording, and both channels will play through the one amp.

Improvements in Fabrication of the Tape

As for the tape itself, old tape was dull on the oxide side with a shiny backing. Newer tape may be textured, or "roughed up" on the non-oxide side to reduce slippage, and may be highly polished on the oxide side to get a smoother surface that will reduce head wear. The tape will be constructed something like that shown in *Figure 24-7*.

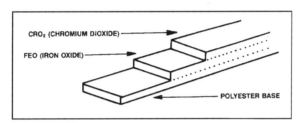

Figure 24-7. Audio tape is constructed something like this.

Every time a tape is recorded in a tape recorder, it is first erased. To cause this erasure, a high level of the ac bias signal is applied to an 'erase' head at the beginning of the tape path. Lower frequency ac (60Hz) can be used over a large length (entire tape) in a "bulk eraser." The tape is either placed on the eraser for a short time, or a handheld bulk eraser is moved around the reel or case a few times.

The same heads are generally used for playback and recording. Some machines, however, have separate play and record heads, primarily to make immediate monitoring of recorded sounds possible.

Chapter 25
Dynamic VCR Head Check
By WG

Testing to determine if a VCR video head is working may be difficult, inconclusive, or both. You can measure the static parameters of the head (inductance, resistance, etc.), and you can observe the picture produced on a TV screen by the VCR playing a known-good tape, but if there are playback problems that make you suspect a bad head there is no way to know if the head itself is functional unless you go to the time, trouble and expense of replacing the head with a new one.

Try This Dynamic Test

A simple and quick technique that you can use to check a head to see if it is actually working is by using a quickly-constructed, hand-made tool. Other equipment that is needed includes a VCR with known-good heads (for experimentation), and RF frequency generator (or a function generator) to generate a signal to be injected, and an oscilloscope and/or a TV to observe the resulting signal. I don't recommend using a TV alone, at least not until you master the technique. Even then, it's best if you have an oscilloscope.

The technique is simple signal injection, using a small inductor (i.e., coil) to radiate a generated signal into the revolving heads (*Figure 25-1* and *Figure 25-2*). A two lead inductor is used to "broadcast" a signal from the signal source.

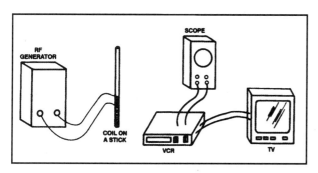

Figure 25-1. By using an inductor to couple a signal from the signal generator to the heads of a VCR while playing a tape, it is possible to determine if the heads are functional.

25 Dynamic VCR Head Check

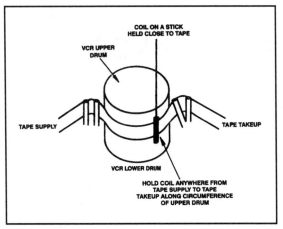

Figure 25-2. To check the VCR heads, connect the coil to the signal generator and hold it near the VCR heads while the VCR is playing a tape.

Setting Up the Equipment

Start by setting up the VCR with a tape that has something recorded on it (one that you don't care about because there is a possibility that it can get damaged), so you can do either or both of the following:

- monitor the FM envelope at the VCR signal with an oscilloscope.
- monitor the VCR output on a TV (i.e., watch the picture generated by a tape being played on a TV screen)

Turn on the signal generator to almost any frequency from around 0.5 MHz to as high as 12 MHz. Take any two-lead coil from your spare parts box and hook it up to the signal generator directly: one lead from generator ground to one leg of the coil, and another lead from the signal output of the generator to the other leg of the coil. The type of signal is not critical, but sine waves give a good picture.

You can experiment with triangle waves, square waves, ramp waves, modulated waves, sweep generated waves, etc., but to describe the technique, this article will use sine waves.

Performing the Check

With the generator on and the output at a medium to maximum setting, carefully place the coil behind the moving tape being played in the VCR, near the revolving heads (upper drum assembly).

Monitor the FM envelope and carefully move the coil closer and farther away from the revolving heads to get the best waveform on the oscilloscope. Try adjusting the RF frequencies anywhere in the range of 0.5MHz to around 2MHz (or even as high as 12MHz). Also adjust the RF generator output level and/or the scope vertical attenuation.

With very little experimentation you should easily get a "marker" blob on the FM envelope or a series of lines and/or washout on the TV screen, at the top, middle or bottom, depending upon the position of the coil relative to the VCR heads. If you try a coil and you get no waveform on the oscilloscope, or indication on the TV screen, try a different one.

Most Inductors Should Work

When I was initially developing this technique I tried several different inductors from my stock of reclaimed parts. Just about all of them worked to push an RF signal out of the coil, into the air, through the VCR tape and into VCR heads where the signal was picked, processed/amplified and passed through the VCR circuits and on to the TV screen. Some of the types and styles of coil that worked are shown in *Figure 25-3*.

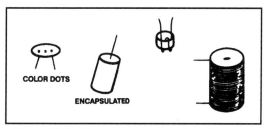

Figure 25-3. Almost any type of coil can be used to perform the dynamic head check.

25   *Dynamic VCR Head Check*

Take Care in Constructing and Using the Coil

Since there is a possibility of catching the coil on revolving parts and causing damage to the VCR, or my own self-made tool I chose a narrow diameter coil about 3/4 of an inch long encapsulated in a smooth, hard, slick coating. I hot-melt glued the coil onto a stick for a handle shown in *Figure 25-4*. As stated before, almost any coil will broadcast enough to be useful when held close enough to, but never touching the tape.

To recapitulate, to check whether the actual heads of a VCR are truly working (assuming the rest of the VCR is functioning) take the cover of the VCR off, hook your "coil on a stick" (*Figure 25-4*) up to any RF signal generator or function generator, connect the VCR to a TV so that you can watch the output on the screen, and/or use an oscilloscope to monitor any appropriate VCR signal (the FM envelope is easy, quick and good).

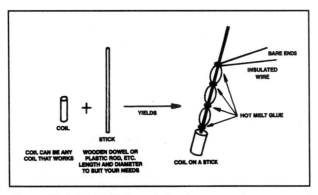

Figure 25-4. The VCR dynamic head tester can be constructed by gluing a coil to a wooden dowel, plastic rod, etc.

By carefully positioning the coil near the revolving heads, between the moving tape and the tape cassette, you can observe the waveform on the oscilloscope or look for trash/washout/lines on the TV screen. That's it.

If the signal/marker shows up on the FM waveform, that means that the head is functioning. If snow/lines/trash shows up on the TV screen, that means that the heads are working (this is a little tricky at first to interpret, so I recommend using an oscilloscope until you become familiar with the particular TV pattern of your setup).

Comments

Although almost any inductor may be used, I would suggest a long smooth one. To work, the inductor has to be held right next to the head, a difficult and tricky task sometimes. Non-cylindrical coils could potentially be caught or grabbed or bind up more easily than a smooth cylindrical one. Moreover, a long inductor (around 5/8 to 1 inch in length) covers the entire width of the tape, ensuring at some point that the video head itself passes next tot the inductor. Hold the inductor near, but not touching the tape, around 1/16 inch, depending on the strength of the RF signal, the scope vertical setting, and your particular coil. If you can't find a smooth coil, heat shrink a piece of heat-shrinkable tubing around a rough coil.

Use the Coil to Check All the Heads

The beauty of this technique is that is allows you to quickly check all heads of any VCR (2-head, 3-head, 7-head, etc.). Remember, some heads work only on certain speeds or in certain modes (i.e., fast forward, freeze frame, frame advance, etc.) All you have to do to check all of the heads is to play a tape that has been recorded at each of the speeds (SP, LP and EP or SLP) on the VCR and at each speed use FF, freeze frame, etc. Then observe the FM envelope at each setting. As an example, say that on one particular VCR setting, one of the heads (only two heads are used at a time) is bad, one of the A or B head waveforms will not show any coil marker (trash) on it (assuming the rest of the circuit is okay).

In another example, both the heads are picking up the RF signal from the coil but the VCR tape picture on the TV screen is full of speckles. Hence, this dynamic technique shows that both heads are functional for the RF signal, but only one head is properly playing on the TV screen. On the FM envelope, one head shows the FM envelope properly and the signal from the other head is almost a flat line (*Figure 25-5*). The greatest likelihood (since both heads pass the RF signal correctly) is that one head is excessively worn. You can confirm this with a head protrusion gauge.

25 Dynamic VCR Head Check

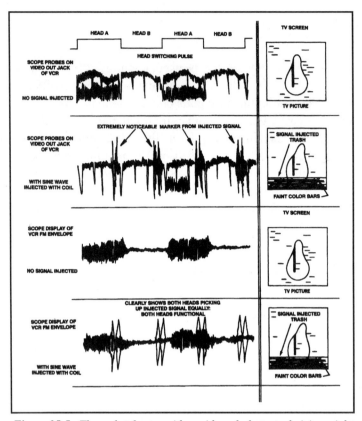

Figure 25-5. These sketches provide an idea of what a technician might expect to see when performing a dynamic check of the VCR heads. One of the heads is not producing a video signal even though it is picking up a signal from the coil. This suggest that the head is functional, but is so worn that it is not making proper contact with the tape.

Experimenting With Frequencies and Positioning

By experimenting with various frequencies you can get horizontal and/or vertical and/or crosshatch bars/lines on the TV screen, or colored bands, etc. On the scope, watching the FM envelope at certain frequencies shows the familiar S curve of FM alignment.

I have also found that the angle at which you position the coil relative to the video head drum doesn't matter very much. You're just trying for an easily generated, quick and dirty signal to enter the heads, not to maximize the effi-

Dynamic VCR Head Check   **25**

ciency with which the signal is coupled to the heads. In most cases you'll find that this RF signal doesn't even leave a trace on the tape when the tape is played back.

This is not a technique in this form or presentation of deriving meaningful waveforms, but simply the presence of an injected waveform at the output confirms that the VCR head is functioning. I suspect that with refinement and appropriate equipment this technique would easily lend itself to checking much more.

The beauty of this technique is that it doesn't require fancy or expensive instrumentation or even knowledge of where to hook up probes (or the sometimes exasperating mechanical difficulties in finding test points and/or attaching probes to them). You simply remove the cover of the VCR, insert a tape, play it and monitor the FM envelope and/or video out jack and/or TV screen while you inject a signal via your coil on a stick.

INDEX

SYMBOLS

16-TOOTH GEAR 54
30HZ REFERENCE SIGNAL 6
8-MM 177, 178, 179
8-TRACK RECORDERS 200
8MM CAMCORDER 1

A

A/D CONVERTERS 34
A/V OUTPUT JACKS 144
AC ADAPTER 78, 82
AC BIAS 196, 201
AC SERVOS 188
ACE HEADS 81
ACETONE 80
ACTUATOR COILS 34, 36
ADJUSTMENT POTENTIOMETER 37
AFC BLOCK 125
AFC CIRCUIT 9, 11, 117, 119, 125
ALCOHOL 80
ALIGNMENT 161, 190, 194, 196, 197
ALPHA WRAP 168
AMPLIFIERS 19, 146, 188
AMPLITUDE 27, 28, 30, 76, 104, 121, 123, 124, 125, 127, 165, 188, 196
ANALOG SAWTOOTH DRIVE 32
ANALOG SWITCH 30
ANALOG VOLTAGE 28
ANTI-STATIC BAR 50
APC CIRCUIT 117
ARMS 67
AUDIO 13, 17, 166, 179, 192, 195, 196
AUDIO CIRCUITS 12, 166
AUDIO DROPOUT DETECTION CIRCUITRY 12
AUDIO HEAD 6, 12, 166, 187
AUDIO HEAD SWITCHING 12
AUDIO HEAD SWITCHING RELAYS 12
AUDIO INPUT CONNECTOR 12
AUDIO INPUT JACK 17
AUDIO PHASE SHIFTER CIRCUIT 12
AUDIO PREAMPLIFIERS 12
AUDIO RECORDING 165, 195
AUDIO SIGNALS 12, 196
AUDIO TRACK 178
AUDIO/CONTROL HEAD 13
AUDIO/CONTROL HEAD ASSEMBLY 143
AUDIO/SERVO HEAD 13
AUTO SPEED SELECT CIRCUIT 118, 119
AUTO TRACKING 28
AUTO TRACKING CIRCUITRY 30
AUTOFOCUS 155
AUTOFOCUS LEVEL 37
AUTOMATIC FREQUENCY CONTROL CIRCUIT 117
AUTOMATIC PHASE CONTROL CIRCUITS 11
AZIMUTH 12, 13, 116, 179, 181
AZIMUTH RECORDING 179, 181
AZIMUTH TRACK 7

B

B/W PICTURE 11
B/W PICTURE CIRCUITS 8
BAND OF STATIC 46
BANDPASS FILTERING 12
BANDPASS FILTERS 185
BANDWIDTH 174
BARBER POLE EFFECT 11
BAV PICTURE 9
BEAT FREQUENCIES 181
BELT FAILURES 5
BELTS 5, 15, 17, 79, 80, 85, 107, 108, 109, 111, 112, 137, 147
BENDING 112
BENT OVER PICTURE 9
BETA 12, 13, 169, 170, 171, 174, 175, 176, 177, 178, 179, 181, 184
BIAS 196, 197
BIAS AC 196
BIAS SIGNAL 195

Index

BINARY DATA 39
BLACK GEAR 67
BLACK LEVEL 37
BLACK SYNC BAR 20
BLANKING 23
BLANKING INTERVAL 22
BLOCK DIAGRAM 27, 73
BLOCK DIAGRAM ANALYSIS 27
BLOCKED VIDEO HEADS 103
BLOWN FUSES 142
BRACKET ASSEMBLY 89
BRAKES 56, 57, 139
BRAKE PADS 56
BRAKING 172
BRIDGING WIRE 64
BROKEN GEAR 54
BRUSHES 173
BURST LEVEL 37
BUSHINGS 139

C

CABLE CONVERTER 144
CABLE TRAP 61
CALIPER 108
CAMCORDER 1, 2, 3, 37, 38, 39, 40, 41, 42, 43, 44, 77, 78, 80, 81, 82, 153, 155, 156, 157, 158, 159, 178
CAMCORDER ELECTRICAL ADJUSTMENT 37, 39
CAMERAS 2, 41, 42, 43, 155
CAP/CYL PHASE COMPARATOR 101
CAPACITORS 42, 43, 64, 88, 134
CAPSTAN 6, 13, 47, 77, 79, 83, 88, 89, 100, 103, 104, 119, 120, 121, 127, 140, 142, 157, 158, 166, 172, 186, 187, 188, 189, 190
CAPSTAN BELT 79, 80
CAPSTAN CIRCUITS 84, 101
CAPSTAN DRIVER IC 88
CAPSTAN FG 104, 187
CAPSTAN FG PULSE 121
CAPSTAN FG SIGNAL 103
CAPSTAN FLYWHEEL 117
CAPSTAN FREE RUN 103
CAPSTAN FREQUENCY GENERATOR 6
CAPSTAN MOTOR 74, 75, 83, 84, 88, 89, 90, 119, 120, 121, 122, 130, 132, 139, 142, 143, 166, 186, 187, 188, 189
CAPSTAN MOTOR CONTROL LINE 121
CAPSTAN MOTOR DRIVE CIRCUIT 118
CAPSTAN MOTOR DRIVER 101
CAPSTAN MOTOR FLYWHEEL 142
CAPSTAN PHASE 7
CAPSTAN PHASE COMPARATOR 101
CAPSTAN PROBLEM 103
CAPSTAN PULLEY 80
CAPSTAN REFERENCE GENERATOR 187
CAPSTAN ROTATION 187
CAPSTAN SENSORS 79
CAPSTAN SERVO 100, 116, 118, 186
CAPSTAN SERVO CIRCUITS 13, 84, 100, 142
CAPSTAN SERVO LOCK 13
CAPSTAN SHAFT 81, 108, 110, 112
CAPSTAN SPEED 119
CAPSTAN/CYLINDER SERVO PROCESS IC 128
CAPSTAN/CYLINDER SERVO PROCESSOR 127
CARRIAGE LATCHING PROBLEMS 152
CARRIAGE LINK GEAR ASSEMBLY 86
CARRIERS 186
CASSETTE 16, 67, 140, 163
CASSETTE CARRIAGE 86
CASSETTE CONTROL SWITCH 87
CASSETTE DOOR HINGE 138
CASSETTE EJECT SWITCH 15
CASSETTE FLAP 109
CASSETTE FORMAT 201
CASSETTE HOUSING 86
CASSETTE LIGHT 65
CASSETTE LOADING MOTORS 67, 189
CASSETTE TEST JIG 137
CASSETTE TRAY 15
CASSETTE-IN SENSOR 67
CASSETTE-LOADING MECHANISM 66, 67, 68
CCD 156
CCD IMAGER 156

Index

CD PLAYERS 37
CENTERING 188
CHAMOIS LEATHER 3, 4
CHAMOIS SWABS 81, 147
CHANNEL CONTROLS 8
CHANNELS 179
CHARACTERISTIC CURVE 196
CHASSIS 17, 97, 137, 161
CHROMA 156, 171, 186
CHROMA FREQUENCIES 178, 181
CHROMA SIGNALS 182, 185
CHROMINANCE CIRCUITS 8
CHROMINANCE SIGNALS 156
CIRCUIT BOARD ASSEMBLY 90, 134
CIRCUIT BOARDS 17, 42, 61, 94
CIRCUIT CARD ASSEMBLIES 5
CIRCUIT TRACE 95
CIRCUITRY 12, 30, 32, 34, 72, 127, 157
CIRCUITS 19, 21
CLEANING 1, 2, 146
CLEANING FLUID 3
CLEANING KIT 3
CLEANING STICK 3, 4
CLEANING TAPE 3
CLEAR/SNOWY PICTURE 7
CLICKS 17
CLIPS 81
CLUTCH TIRES 16, 17
CMOS DEVICES 146
COILS 186, 203, 205, 206, 207, 209
COLLECTOR 72
COLLECTOR PIN 67
COLOR 10, 11
COLOR CHANGE 11
COLOR CIRCUITS 8
COLOR CONTROL CIRCUIT 11
COLOR FREQUENCY 184
COLOR MONITOR 39
COLOR SEQUENCER CIRCUITS 184
COLOR SUBCARRIER FREQUENCY 11
COLOR/ LUMINANCE CIRCUITS 9
COMPARATORS 101, 103
COMPOSITE VIDEO SIGNAL 194

CONCENTRIC 57
CONDUCTOR 198
CONTAMINANTS 3
CONTINUITY 84
CONTROL BOARD 157, 159
CONTROL CIRCUITRY 28, 30
CONTROL IC 104
CONTROL PC BOARD 80
CONTROL SIGNAL 117
CONTROL TRACK 6, 9, 99, 118
CONTROL TRACK LOGIC PULSES 6, 9
CONTROLLER 189, 190, 191
CONVERSION CIRCUIT 8
CONVERTER 61
COUNT PULSES 24
COUNTER BELT 14
COUNTER CIRCUIT 14
COUNTING PULSES 21, 24
CRACKED LOADING BELT 86
CRACKS 43
CRISPENED COMB FILTER 193
CROSSHATCH BARS 208
CROSSHATCH LINES 208
CROSSTALK 172, 174, 179, 181, 182
CTL 99, 100, 101, 104, 117
CTL AMP 103
CTL FREQUENCY 119
CTL GEN 103
CTL HEAD 103, 118
CTL HEAD CIRCUIT 103
CTL PULSE 118, 119, 121
CTL SIGNAL 103
CYL/CAP PHASE COMPARATOR 101, 103
CYLINDER 3, 6, 12, 73, 81, 84, 89, 90, 99, 100, 103, 104, 118, 127, 130, 131, 132, 157, 158, 168, 170, 171, 172, 173, 174, 186, 187, 188, 189, 190
CYLINDER ASSEMBLY 90
CYLINDER DRIVE CIRCUITS 134
CYLINDER ERROR LINE 131
CYLINDER FG SIGNAL 101
CYLINDER FG/PG 104

Index

CYLINDER FREE RUN 103
CYLINDER HEAD MOTOR 163
CYLINDER MOTOR 74, 87, 89, 90, 91, 101, 127, 130, 131, 132, 133, 134, 186, 188
CYLINDER MOTOR DRIVE IC 131
CYLINDER MOTOR SERVO 101
CYLINDER MOTORS 87
CYLINDER PG PULSES 101
CYLINDER PULSE GENERATOR 6
CYLINDER PULSE GENERATOR SIGNAL 6
CYLINDER PWMS 104
CYLINDER ROTATION 15, 187
CYLINDER SERVO 100
CYLINDER SERVO INTERFACE IC 127, 133
CYLINDER SPEED COMPARATOR 101

D

D/A CONVERTERS 32
DC VALUE 118
DC/DC CONVERTER 61, 64, 158
DEAD UNIT 109
DEAD VCR 112
DEFECTIVE COIL 63
DELAY 25, 30
DELAY DEVICE 183
DELAY LINE 9
DELAY LINE CIRCUITRY 9
DELTA TIME 22, 24
DELTA TIME FUNCTION 19, 24
DELTA TIME METHOD 19
DEMODULATION 174
DEMODULATOR 192
DENTS 42
DETECTOR 138, 191
DEW SENSOR 54, 156, 190
DEW SENSOR CIRCUIT 14
DIAGNOSIS 72
DIAGONAL 27
DIAGONAL TRACKS 116
DIGITAL ADJUSTMENT TOOLS 39
DIGITAL ADJUSTMENTS 37
DIGITAL AUDIO RECORDING 153
DIGITAL CIRCUITRY 37
DIGITAL DATA 39
DIGITAL NATURE 37
DIGITAL TO ANALOG CONVERTER 38
DIGITAL TRANSISTOR 62
DIGITALLY ENCODED MUSIC 37
DIMMER 61
DIODES 62, 90, 131
DIRTY 1, 12
DIRTY CAPSTAN 103
DIRTY CTL HEAD 103
DIRTY DRIVE BELTS 103
DIRTY HEADS 9
DIRTY VIDEO HEADS 103
DISPLAY CIRCUIT BOARD 61
DISTINCT HORIZONTAL LINES 46
DISTORTED VIDEO 46
DISTRIBUTOR 3
DMM 121, 141
DOC 172
DOWN-CONVERTED CHROMA 13
DRIVE AMPLIFIERS 32
DRIVE BELTS 83, 85, 86, 103
DRIVE CIRCUITS 188
DRIVE CONTROL LINE 121
DRIVE GEARS 56
DRIVE GENERATORS 30
DRIVE MOTOR 138
DRIVE SIGNALS 27, 28, 30, 32, 35
DRIVER IC 84, 85, 88, 91, 127
DRIVER MOTOR 91
DROPOUT COMPENSATION CIRCUIT 11
DROPOUT COMPENSATOR 172
DROPOUTS 172
DRUM 27, 47, 49, 51, 84, 90, 116, 123, 170
DRUM ASSEMBLY 32, 34, 127
DRUM CIRCUIT BOARD ASSEMBLY 90
DRUM CYLINDER 90
DRUM CYLINDER PG COIL 90
DRUM MOTOR 84, 89, 90, 124
DRUM MOTOR CIRCUITS 84
DRUM SERVO 100, 118, 120, 127, 186

Index

DRUM SERVO PROBLEM 103
DUAL TRACE 23
DUAL-TRACE SCOPE 22
DUTY CYCLE 118, 132
DVM 39

E

E-CLIP 163
EAR ASSEMBLY 85
EATING OF THE TAPE 108, 109
ED BETA 177, 178
EEPROM 38, 39
EIA 5
EJECT 65
ELECTRICAL ADJUSTMENTS 188
ELECTRICAL CAMCORDER ADJUSTMENT 40
ELECTRICAL CIRCUITRY 82
ELECTROLYTIC CAPACITORS 62, 88, 90
ELECTROMAGNETIC HEAD 195
ELECTROMAGNETS 45, 197
ELECTRON BEAM 20
ELECTRONIC SWITCH 19
ELECTRONICS 44
ELECTRONICS INDUSTRIES ASSOCIATION 5
ELECTRONICS THEORY 194
EMITTER 67, 72
ENCODER CIRCUIT BOARD 158
END OF TAPE SENSOR LAMP 16
END OF TAPE SENSORS 14, 15, 67, 191
ENVELOPE DETECTOR 30
ENVELOPE DETECTOR CONTROLS 28
EOT SENSORS 191
EP HEADS 8
EP MODE 8, 181, 193, 207
EQUALIZATION 200
EQUALIZING PULSES 22
ERASE HEAD 194
ERASER 201
ERROR VOLTAGES 129
ETHYL ALCOHOL 4
EXCESSIVE DROP-OUT 46

F

F-CON 144, 145
FAST FORWARD 16, 46, 56, 86, 90, 109, 119, 137, 163, 207
FAST REVERSE 161, 163
FC 11
FEED REEL 139
FEEDBACK PULSES 76, 130
FEEDBACK SIGNALS 76, 127
FELT CLUTCH 15
FELT PADS 56, 139, 147
FERROMAGNETIC 196
FF TORQUE 150
FG FREQUENCY 117
FG HEAD 121, 122, 125
FG PULSE FREQUENCY 119
FG PULSE TRAIN 117
FG PULSES 117, 119, 122, 123, 125, 127, 187
FG SIGNAL 103
FILTERS 13, 185, 198
FINE TUNING CHANNEL CONTROLS 8
FISHER 17
FLAGGING 9
FLAT-SCREEN TV 153
FLOPPY DISK 39
FLOW PATHS 74
FLUORESCENT LAMP 50
FLUORESCENT LIGHT 6, 7, 75, 130
FLYING ERASE HEADS 32
FLYWHEEL 80
FM 171, 185
FM ALIGNMENT 208
FM CARRIER 165, 166, 167, 171, 173, 178
FM DEMODULATOR 9, 167, 173, 185, 186
FM ENVELOPE 204, 205, 206, 207, 209
FM FREQUENCIES 174, 179
FM INTERLEAVING 181
FM LIMITERS 9
FM LUMINANCE 13, 181, 184, 185
FM LUMINANCE FREQUENCY 12

215

Index

FM MODULATION 174
FM MODULATOR 167, 185
FM SIGNALS 12, 13, 28, 165, 180
FM SOUND DETECTOR 192
FOCUS 82, 155
FOREIGN OBJECTS 16, 141
FOUR HEADS 193
FRACTURES 42
FRAME ADVANCE 207
FREEZE FRAME 207
FREQUENCIES 6, 104, 165, 174, 198, 200
FREQUENCY COUNTER 39
FREQUENCY GENERATOR 117, 187
FREQUENCY GENERATOR PULSE 187
FREQUENCY TRANSLATION 165
FRINGING 199
FUNCTION GENERATOR 203
FUSE 62, 88, 112, 113, 142, 157, 159
FUSIBLE RESISTOR 87

G

GAP 180
GARBLED AUDIO 13
GAUZE 4
GE 17
GEAR ASSEMBLY 54, 86, 161
GEAR-GRINDING 53
GEARS 94, 111, 163
GENERATOR 186, 187, 198, 204
GERMANIUM TRANSISTOR 72
GOLDSTAR 93
GREASE 80, 82, 147
GUARD BANDS 7, 116, 179
GUIDE HEIGHTS 150, 152
GUIDE ROLLERS 81
GUIDEPOST TRACKS 109
GUIDEPOSTS 49, 109, 140, 143, 144
GUIDES 81

H

HALF-OMEGA WRAP 168
HALL DEVICE 127, 133
HALL EFFECT GENERATOR CIRCUITS 89
HALL EFFECT SENSORS 14
HALL-EFFECT DEVICE 127
HEAD 27, 116, 168, 198, 199, 201, 203, 207
HEAD AMP SELECT CIRCUIT 101
HEAD AMPLIFIER 19
HEAD AMPS 101, 104
HEAD CLEANING KIT 3
HEAD CLOGGING 5
HEAD COIL 117
HEAD CYLINDER 174, 187, 194
HEAD DRUM 3, 27, 118, 121, 127, 142
HEAD DRUM PHASE CONTROL 118
HEAD DRUM SPEED CONTROL 118
HEAD GAP 116, 179, 198
HEAD GAP WIDTH 171
HEAD NONLINEARITY 198
HEAD PREAMP 185
HEAD PROTRUSION GAUGE 207
HEAD SWITCH CONTROL 24
HEAD SWITCH GENERATOR 128, 129
HEAD SWITCHING 19, 21, 23, 24, 101, 102
HEAD SWITCHING CIRCUITS 8, 20, 28, 185
HEAD SWITCHING POSITION 103
HEAD SWITCHING RELAY 12
HEAD SWITCHING SIGNAL 19, 35, 101
HEAD WEAR 46
HEAD-TO-TAPE CONTACT 172
HEADS 3, 7, 8, 172, 173, 179, 184, 185, 193, 207
HEAT SINK FINS 64
HELICAL PATH 116
HELICAL SCAN 167
HELICAL SCAN METHOD 116
HELICAL SCAN RECORDING 116
HELICAL TRACKS 168
HELIX 168
HERRINGBONE PATTERN 9
HEXADECIMAL DIGITS 39
HEXADECIMAL REPRESENTATION 39
HI-FI 11, 12, 178

Index

HI-FI AUDIO 11, 12, 13, 46
HI-FI AUDIO HEADS 12
HI-FI HEADS 19, 30
HI-FI RECORDINGS 166
HIGH BAND 167
HIGHER FREQUENCIES 196, 199
HIGHER LUMINANCE 179
HISSING AUDIO 17
HITACHI 1, 61, 77, 78, 79, 161, 162, 163
HITACHI VM-2400 155, 156
HOOK 57
HORIZONTAL LINES 20, 22, 181, 182
HORIZONTAL POSITION 23
HORIZONTAL SCAN LINE 9
HORIZONTAL SYNC PULSES 11, 22, 181
HORIZONTAL TEAR 20
HORIZONTAL VERNIER CONTROL 23
HOWARD SAMS 82
HQ CIRCUIT 193
HUB LOCK MECHANISMS 16
HYBRID ICS 188
HYSTERESIS LOOP 197

I

IC TRANSFORMER 62
ICS 5, 83, 89, 104, 120, 146, 185
IDLER ASSEMBLY 16, 17, 139, 140
IDLER GEAR 94
IDLER TIRES 56, 107, 108, 109, 110, 111, 140
IDLER WHEEL 56, 137, 139, 140
IDLER/TAKE-UP SYSTEM 140
IDLERS 5, 17, 110, 140, 147
IMAGE SENSOR 156
IMPEDANCE ROLLER 81
INCANDESCENT LAMPS 138
INCOMPLETE LOADING 108, 109
INDUCTANCE 203
INDUCTOR 203, 207
INNER ARM 67
INSPECTION 2, 3
INTEGRATOR OUTPUTS 105
INTEGRATORS 101, 104, 120

INTERLACED SYNC 22
INTERMITTENT MODE SWITCH 161
INTERMITTENT SHUTOFF 109, 111
INTERMITTENT SOUND 13
INTERVAL 25
IR DETECTOR 146
IR LED 146
IR TAPE LOADING SENSORS 66
ISOPROPYL ALCOHOL 81

J

JIGS 82, 137, 138
JITTER 6, 103, 112, 173, 174, 184
JITTER COMPONENTS 184
JITTERY VIDEO 13
JOG/SHUTTLE MODES 27, 34
JUMPING 102

L

LAMP 16
LATCHING MECHANISM 56, 57
LEAF SWITCHES 138, 141
LED 138, 190, 191
LENS 42, 43, 82
LENS ASSEMBLY 80
LENS CLEANING SOLUTION 82
LENS EXTENSION 42
LENS TISSUE 82
LENSES 42
LEVERS 67
LIGHT METER 39
LIGHTING 39, 67
LINEAR TRACK 99
LINK GEAR 86
LINT-FREE GAUZE 3
LOAD MOTOR 141
LOAD MOTOR PULLEY 57
LOADING BELTS 86, 111
LOADING GEARS 141
LOADING MECHANISM 5
LOADING MOTOR 85, 86, 89, 109, 111
LOADING MOTOR DRIVE IC 85, 86, 87
LOADING MOTOR FORWARD 73

217

Index

LOADING MOTOR IC 87
LOADING MOTOR MODE CAM ASSEMBLY 86
LOOPS 104, 121, 131, 186
LOSS OF SYNC 99
LOW BAND 167
LOW VOLTAGES 17
LOW PASS FILTER 131, 133, 134
LOWER CYLINDER 185
LP 181, 207
LP SPEED 181
LUBRICATION 1, 2, 4, 146, 147
LUMA 171
LUMA/CHROMA CIRCUITS 129
LUMINANCE 156, 171, 173, 174, 175, 176, 177, 178, 183, 185, 186
LUMINANCE CIRCUITS 8, 185
LUMINANCE FM 181
LUMINANCE FM FREQUENCIES 178

M

MAGNAVOX 17, 127
MAGNETIC FIELD 185, 195, 197, 198
MAGNETIC HEAD 195
MAGNETIC RECORDING 165, 166, 195
MAGNETIC RECORDING HEAD 197
MAGNETIC RECORDING TAPE 195
MAGNETIC TAPE 198, 200
MAGNETIC TAPE RECORDING 116
MAGNETIZATION 197
MAGNETIZING FIELD 196
MAGNETS 186
MAGNIFYING GLASS 47, 64
MAIN CIRCUIT BOARD 54
MAIN GEAR 55
MAINTENANCE 1, 2, 3
MANUFACTURERS 3, 39
MASTER CAM GEAR 86
MATRIXED 118
MATSUSHITA 17
MECHACON 190
MECHACON SWITCH 190
MECHANICAL ALIGNMENT 143
MECHANICAL POSITION SWITCH 88

MECHANICAL POTENTIOMETERS 39
MECHANICS 44
METAL BAND 172
METAL SUPPORT 95
METER 194
MICROPROCESSOR 37, 38, 67, 72, 74, 79, 84, 90, 137, 163, 188
MISSING TOOTH 53
MITSUBISHI 27
MODE CAM 77
MODE SWITCH 10, 12, 13, 15
MODULATED WAVES 204
MODULATOR 192
MONOCHROME VIDEO 181
MOTOR 15, 84, 94, 142, 158
MOTOR BRUSHES 93
MOTOR CONTINUITY 84
MOTOR CONTROL CIRCUITS 142
MOTOR CONTROL IC 86
MOTOR CONTROL LINE 131
MOTOR DRIVE 188
MOTOR DRIVE AMPLIFIERS 188
MOTOR DRIVER IC 87, 88, 133, 142
MOTOR DRIVERS 104
MOTOR FUSE 83
MOTOR LEADS 84
MOTOR PLUG 90
MOTOR PULLEY 54, 80
MOTOR SOCKET 90
MOTOR SOCKET TERMINALS 88
MOTOR SUPPLY CIRCUITS 84
MOTOR WINDINGS 84, 88
MOTORBOATING SOUNDS 17
MOTORS 5, 16, 104, 188, 189
MOUNTING SUPPORT 95
MOUNTING TAB 95
MOVABLE HEAD MODE 27
MOVABLE HEADS 35
MOVABLE HEADS DRIVE CIRCUITRY 27
MULTITRACK RECORDINGS 200
MV ACTUATOR COILS 32
MV DRIVE AMPLIFIER CIRCUITRY 32, 35
MV DRIVE CIRCUITRY 35

Index

MV DRIVE GENERATORS 30, 35
MV DRIVE SIGNAL OFFSET RANGE 35
MV DRIVE SIGNALS 30, 32, 35
MV HEAD ACTUATOR COILS 34
MV HEAD DRIVE CIRCUITRY 34
MV HEAD DRIVE GENERATORS 28
MV HEADS 27, 30, 32, 34, 35, 36
MV HEADS ACTUATOR COILS 32, 34
MV MODE 34

N

NATIONAL TELEVISION SYSTEM COMMITTEE 99
NOISE 9, 17, 20, 104
NOISE BANDS 99, 103, 193
NOISE BARS 6, 13, 27, 35, 143
NOISE CANCELLER CIRCUIT 9
NOISE-FREE CUE 194
NOISE-FREE PAUSE 194
NOISY 12
NOISY PICTURE 8, 13
NON-HI-FI 6, 12
NON-HI-FI AUDIO 6
NON-INTERLACED SYNC 22
NPN TRANSISTOR 72
NTSC 99

O

OFF-PITCH AUDIO 13
OFFSET ADJUSTMENT 35
OFFSET VOLTAGE 35
OHMMETER 62, 84
OIL 4, 147
OILING 4
OMEGA WRAP 168
ONE-TOUCH RECORDING 17
OPERATIONAL AMPLIFIERS 34, 120
OPTICS 44, 155
OSCILLATOR FREQUENCY 11
OSCILLATORS 8, 34, 117, 171
OSCILLOSCOPE 19, 21, 22, 24, 35, 39, 121, 127, 194, 203, 204, 205, 206

OUT OF SYNC PICTURE 13
OUTPUT PORTS 74
OXIDE 95, 201
OXIDE COATING 110, 199

P

PG COIL 90
PANASONIC 15, 17
PAPER INSULATOR 97
PARTIAL PICTURE 46
PARTICLES 43
PAUSE FUNCTION 46, 47
PB FM SIGNAL 104
PC BOARD 72, 78, 80, 81, 82, 120, 121, 157
PCB 185, 188, 189
PEAK WHITE FREQUENCIES 176, 177, 178
PERIPHERAL COMPONENTS 104
PG PULSE FREQUENCIES 123, 134
PG PULSES 101, 123, 127, 128
PG SHIFTER 104
PG/FG GENERATOR BLOCK 127
PG/FG SEPARATOR 128
PG/FG SEPARATOR BLOCK 127
PHASE COMPARATORS 101, 129
PHASE CONTROL LOOP 117
PHASE GENERATOR HEADS 118
PHILIPS 82
PHOTO DETECTOR 190
PHOTO DETECTOR ASSEMBLY 138
PHOTODETECTOR 138, 191
PHOTOTRANSISTOR 67, 191
PICKUP COILS 186
PICTURE 1, 6, 8, 9, 10, 13, 30, 38, 45, 46, 49, 75, 84, 99, 102, 103, 161, 169, 178, 193, 204, 207
PICTURE PULSATES 13
PICTURE ROLL 112
PINCH ROLLERS 47, 81, 103, 107, 108, 110, 111, 112, 140, 190
PINS 57, 63, 95
PLASTIC FOAM 81
PLASTIC MOUNTS 43

219

Index

PLAY MODE 66, 90, 99, 119, 129, 131, 134, 137, 142, 161, 175, 177, 192, 194
PLAY ROLLER 15
PLAYBACK 6, 9, 10, 11, 13, 23, 28, 34, 38, 47, 74, 76, 100, 101, 103, 104, 116, 144, 165, 166, 167, 182, 184, 185, 186
PLAYBACK HEADS 11, 47
PLAYBACK SIGNALS 19, 184
PLAYBACK/RECORD MECHANISM 110
PLAYBACK/RECORD SWITCHING TRANSISTOR 8
PLUG 90
PNP TRANSISTOR 72
POLARITY 183
POOR TRACKING 112
POTENTIOMETERS 7, 115
POWER CIRCUIT 159
POWER SUPPLY 8, 16
POWER SUPPLY CIRCUITS 104
PRE-PROCESS CIRCUITS 156
PRE-PROCESS IC 156
PREAMPLIFIERS 8, 28, 185
PRESSURE PLATES 139
PRINTED CIRCUIT BOARDS 50, 68
PRINTED CIRCUIT BOARD TRACE 163
PRINTTHROUGH 200
PROJECTION TV 153
PROTECT CIRCUITRY 34
PULLEYS 5, 47, 80, 112
PULSE GENERATORS 173, 186
PULSE TRAIN 117
PULSE WIDTH MODULATION SIGNAL 6
PULSE WIDTH MODULATOR 100, 117
PULSES 127, 186
PWM 101, 117, 118, 119, 130, 131, 132, 133
PWM BLOCKS 118, 120, 129, 132, 133, 134
PWM OUTPUTS 105, 132, 133, 134

Q

Q-TIPS 82
QUAD RECORDER 171
QUASI-TRAPEZOIDAL WAVE 123

R

R/S/F 104
RACKING CONTROL 14
RADIO SHACK 77
RAM 38
RAMP WAVES 204
RASPY 12
RCA 15, 77
RCA VLP900 90
RCA VLT650HF 89
READ ONLY MEMORY 38
RECORD 9, 17, 23, 74, 75, 76, 82, 99, 100, 102, 103, 104, 116, 165, 166, 167, 169, 177, 194
RECORD HEADS 11
RECORD TAB 16, 17
RECORD TAB MICROSWITCH 16
RECORD TAB SWITCH 15
RECORD/PLAYBACK RELAY 12
RECORDED STRIPE 27
RECORDED WAVELENGTH 198
RECORDING CIRCUITS 21
REEL 138, 143
REEL DRIVE MOTORS 189
REEL HEIGHT GAUGES 143
REEL HEIGHTS 152
REEL MOTION SENSORS 138, 139
REEL SENSORS 14, 190, 191
REEL SPINDLES 103
REEL TABLE 139
REEL TABLE HEIGHTS 150
REEL-TO-REEL 168
REEL-TO-REEL TYPE RECORDERS 168
REELS 14, 17
REFERENCE GENERATORS 186
REFERENCE SIGNALS 8, 9, 76, 121
REGULATOR 157
RELAY CIRCUITRY 32
RELAYS 5, 32, 34
REMOTE CONTROL 146
REMOTE UNIT 5

Index

RESISTANCE 203
RESISTORS 42, 90, 124
REVERSE 119
REVOLVING HEADS 205
REWIND 16, 56, 87, 90, 109, 110, 137, 142
REWIND MODES 137
REWIND SENSOR 15
REWIND TORQUE 150
RF CHANNEL 144
RF CONVERTER 12
RF ENVELOPE 152
RF FREQUENCY GENERATOR 203
RF GENERATOR 205
RF INPUT 192
RF MODULATOR 8, 144, 145, 186
RF SIGNAL 205, 207, 209
RF SWITCHING 192
RIPPLE 104
ROLLING 6
ROM 38
ROTARY TRANSFORMER 8, 12, 173, 184
ROTARY TRANSFORMER WINDING 184
ROTATING CYLINDER 173
ROTATING DRUM 50
ROTATING WINDING 184
ROTOR 184
RUBBER 107
RUBBER PARTS 107, 113
RUBBER REVITALIZER 17
RYDER PRESS 82

S

S CURVE 208
S-VHS 177
SANYO VHR9300 53
SAWTOOTH 35
SCANNER 166, 167, 170
SCANNER MOTOR 166
SCANNING LINE 181
SCHMITT TRIGGER 100
SCRATCHES 42
SEARS 77
SELECTION RELAY 32

SEMICONDUCTOR FUSE 62, 64
SENSE LED 79
SENSORS 94, 155, 156, 191
SENSOR CIRCUITRY 17
SENSOR DRIVER 156
SENSOR PCB 156
SENSORS 16, 156, 190, 191
SERVICE LITERATURE 21, 22, 141, 142
SERVICE TECHNICIAN 83
SERVO 6, 13, 75, 84, 100, 102, 103, 116, 142, 143, 163, 166, 173, 188
SERVO BOARD 88
SERVO CIRCUITS 6, 30, 19, 72, 74, 76, 84, 99, 102, 103, 115, 121, 123, 125, 126, 127, 143, 185, 186, 187, 188
SERVO CONTROL 166
SERVO CONTROL CIRCUIT 74, 75
SERVO CONTROL IC 87, 104
SERVO CONTROL VOLTAGE 188
SERVO HEAD 13
SERVO ICS 87, 88, 89, 90, 120, 121, 122
SERVO LOCKS 6, 15, 76
SERVO PROBLEMS 75, 142
SERVO SECTION 189
SERVO SYSTEM 75, 99, 100, 104, 143
SET SCREW 49
SHAPER CIRCUIT 102
SHUT-DOWN FUNCTION 133
SHUTDOWN 53
SIGNAL DETECTION CIRCUITRY 28
SIGNAL FREQUENCY 198
SIGNAL GENERATOR 204
SIGNAL-TO-NOISE RATIO 177
SILICON TRANSISTOR 72
SINE WAVES 204
SINGLE IC 185
SINGLE WIPER 37
SIX HEAD VCR 47
SKEWED PICTURE 9
SLANT TRACKS 169
SLIP RINGS 173
SLIP-CLUTCH 140

221

Index

SLIP-CLUTCH MECHANISM 139
SLOW MOTION 46
SLOW REWIND PROBLEMS 5
SLP 181, 193, 207
SMEARED 38
SNOW 9
SNOW BAR 7
SNOWY PICTURE 17, 46
SOCKET 90
SOFTWARE 39
SOLID LEADS 50
SOUND ABSENT 13
SOUND DETECTOR 192
SOUND GARBLED 13
SP 8, 121, 193, 194, 207
SP HEADS 8, 193
SPARK 95
SPARK DISCHARGE 95
SPARKING 93
SPEED COMPARATOR 101
SPEED CONTROL 173
SPEED CONTROL LOOP 117
SPEED OF ROTATION 50
SPIKE 17, 124
SPINDLE 78
SPINNING HEAD 50
SPRINGS 57, 139
SQUARE WAVES 23, 24, 131, 132, 134, 204
SQUEALING SOUNDS 109, 110, 111
STAGE 66
STAGE LOADING MOTOR 68
STAGE MECHANISM 65, 66, 67
STAGE MOTOR 65, 67
STAGE LOADING MECHANISM 67
STATE SWITCH 16
STATIONARY AUDIO/SERVO HEAD 12
STATIONARY HEAD 118
STEREO FORMAT 201
STEREO RECORDING 201
STOP MODE 10, 12, 15, 175, 191
STROBE 6
STROBE METHOD 130

STROBOSCOPIC EFFECT 130
SUBCHASSIS 142
SUM AMPS 101, 104
SUPER VHS 176
SUPPLY REEL 47, 109, 110, 172, 190
SURFACE ACOUSTIC WAVE 183
SWEEP GENERATED WAVES 204
SWITCHES 5, 67, 94, 190
SWITCHING TRANSISTOR 62
SYMPTOM ANALYSIS 6
SYNC 22, 103
SYNC INTERVAL 20
SYNC LEVEL 37
SYNC PULSES 19, 21, 23, 24
SYNC SEPARATOR 102
SYNC TIP FREQUENCIES 176, 177, 178
SYSCON 146, 188, 189
SYSTEM CONTROL 53
SYSTEM CONTROL CIRCUITS 16, 72, 73
SYSTEM CONTROL IC 73
SYSTEM CONTROL VCO 104
SYSTEM CONTROLLER 186, 188, 189

T

TAKE-UP REEL 109, 110, 111, 139, 140, 163, 168, 190
TAKE-UP REEL END SENSOR 94
TAKE-UP SENSORS 79
TAKE-UP TORQUE 150
TAPE 9, 17, 35, 86, 87, 199
TAPE BACK TENSION 9
TAPE END SENSOR 95
TAPE HEADS 5, 84, 165
TAPE HOLDING COMPARTMENT 141
TAPE INTERCHANGE PROBLEMS 14
TAPE LIBRARY 177
TAPE LOADING 83
TAPE LOADING GEARS 15
TAPE LOADING MECHANISM 53
TAPE LOADING MOTOR 15
TAPE MECHANISM 77, 80
TAPE OXIDE DEFECTS 9

Index

TAPE PATH GEOMETRY 82
TAPE SENSORS 94, 191
TAPE SLACK SENSOR 14
TAPE SLIPPAGE 108, 110
TAPE SPEED 48, 186, 198
TAPE STRETCH 172
TAPE TENSION 7, 48, 49, 172
TAPE TENSION GAUGE 139
TAPE THREADING MOTORS 189
TAPE TRANSPORT 112, 116, 137, 191
TAPE TRANSPORT MECHANISM 93, 94, 107
TAPE TRANSPORT SPEED 100
TAPE TRANSPORT SYSTEM 139
TAPE WIND 77
TAPE WRAP 168
TAPE-EDGE DAMAGE 109, 112
TAPE-WIND BELT 80
TAPES 14, 77
TECHNICAL MANUALS 194
TELEVISION THEORY 194
TENSION BAND 47
TENSION REGULATOR 172
TERMINAL LEADS 84
TERMINAL PIN 63
TEST JIG 82, 137, 138, 139, 144, 147
THOMSON CONSUMER ELECTRONICS 39, 70
THREADED 174
THREADING 43
THREE-HEAD VCR 194
THREE-POSITION SWITCH 115
THUMBWHEEL 8
THUMBWHEEL SETTING 8
TIME DELAY 19, 25
TIMEBASE-FREQ 23
TIMEBASE-FREQ CONTROL 23
TIMER CONTROL 17
TIMING 6, 21
TIMING GEARS 86
TIRE KIT 140
TIRES 140, 147
TORQUE 56, 78, 139, 140, 150, 152
TORQUE CONTROL BLOCK 129
TORQUE GAUGE 139

TOTAL PICTURE LOSS 46
TRACES 43, 95, 163, 194
TRACK 179
TRACK LENGTH 186
TRACK PULSES 187
TRACKING CIRCUIT 6
TRACKING CONTROL 7, 104
TRACKING CONTROL CIRCUIT 9
TRACKING MULTIVIBRATOR 101
TRACKING SET 103
TRACKING SYSTEM 118
TRACKS 109, 140
TRANSFORMER 63, 185
TRANSFORMER WINDING 64
TRANSISTOR BUFFER AMPLIFIER 131
TRANSISTORS 62, 72, 88, 90, 123, 188
TRANSMITTER 189
TRANSPORT ASSEMBLY 95
TRANSPORT SYSTEM 139
TRIANGLE WAVES 204
TRIGGER MODE 23
TRIGGER POLARITY 23
TRIGGER SOURCE 23
TRIPOD 39
TROUBLESHOOTING 5, 6, 71, 103, 104, 115, 135, 159, 163, 194
TUNER 192
TUNER/IF 10, 12
TV 9, 10, 12, 171, 203, 204, 205
TV AFC CIRCUIT 9
TV MODE 10
TV SCREEN 205
TV SETS 171
TV/VCR RELAY 12
TV/VCR SWITCH 10
TWEEZERS 64
TWO DRIVE SIGNALS 36
TWO HEAD VCR 48

U

UHF-VHF TUNING CIRCUIT BOARD 61
UNSTABLE IMAGE 46
UPPER DRUM 45

223

Index

UPPER DRUM ASSEMBLY 36, 205
UPPER VIDEO DRUM 45

V

V SYNC 100, 101, 102, 104
V-LOCK GENERATOR 101, 102
V-SYNC FREQUENCY 104
VACUUM FLUORESCENT DISPLAY 56, 145
VARIABLE TEMPERATURE DESOLDERING TOOL 43
VARIABLE TEMPERATURE SOLDERING TOOL 43
VCO FREQUENCY 11
VCR 4, 5, 6, 9, 10, 11, 12, 13, 14, 15, 16, 17, 19, 20, 21, 22, 23, 25, 27, 28, 34, 45, 46, 47, 48, 49, 51, 53, 54, 61, 62, 65, 66, 68, 70, 71, 72, 73, 74, 75, 76, 77, 78, 83, 84, 85, 86, 87, 88, 89, 90, 91, 93, 94, 95, 97, 99, 103, 105, 107, 108, 109, 110, 111, 112, 113, 116, 119, 121, 123, 129, 131, 134, 137, 138, 139, 141, 142, 144, 146, 147, 149, 150, 151, 152, 153, 155, 156, 157, 161, 162, 163, 166, 167, 168, 169, 171, 172, 173, 174, 175, 176, 178, 181, 185, 186, 188, 189, 190, 191, 192, 194, 203, 204, 205, 206, 207, 209
VCR CASSETTE SENSOR SWITCH 14
VCR CIRCUITS 10, 17, 48
VCR HEAD DRUM 51
VCR HEADS 205
VCR MICROPROCESSOR 67
VCR OIL KIT 3
VCR PINCH ROLLERS 108
VCR SERVO 115
VCR SERVO CIRCUITS 105
VCR SERVO SYSTEMS 127, 135
VCR SHORT CIRCUIT 93
VCR SIGNAL 204, 206
VCR TAPE 205
VCR TAPE TRANSPORT CHASSIS 68
VCR TECHNICIANS 19, 78
VCR/TV 192
VECTORS 182
VECTORSCOPE 39
VERTICAL 27
VERTICAL BLANKING 23
VERTICAL BLANKING PULSES 22
VERTICAL DEFLECTION 20
VERTICAL HOLD 20
VERTICAL JITTER 102
VERTICAL SIGNAL 102
VERTICAL SYNC 20, 22, 76, 101, 102, 187
VERTICAL SYNC PULSES 23, 24, 25, 173
VHS 153, 169, 170, 171, 174, 175, 176, 177, 178, 179, 181, 184
VHS FORMAT 99, 101, 153, 166
VHS TAPE CASSETTE 77
VIDEO 8, 12, 17, 75, 82, 99, 103, 119, 165, 166, 171, 179, 192, 195
VIDEO CAMERAS 41
VIDEO CASSETTES 16, 170
VIDEO CYLINDER 7
VIDEO CYLINDER MOTOR 76
VIDEO DETECTOR 192
VIDEO DRUM 47, 48, 50, 109
VIDEO FM 13
VIDEO HEAD ASSEMBLY 75
VIDEO HEAD DEFECTS 46
VIDEO HEAD DRUM 50
VIDEO HEAD PREAMPLIFIERS 8
VIDEO HEAD PULLERS 51
VIDEO HEAD TESTER 48
VIDEO HEAD TIP PROTRUSION 150
VIDEO HEADS 3, 6, 7, 8, 11, 12, 13, 19, 21, 27, 45, 46, 47, 48, 49, 50, 51, 75, 81, 84, 103, 116, 117, 151, 156, 166, 167, 168, 172, 179, 186, 193, 207
VIDEO IF AMPS 192
VIDEO INPUT CONNECTOR 10
VIDEO LOCK 132

Index

VIDEO OUT JACK 209
VIDEO PREAMP CIRCUIT 9
VIDEO RECORDERS 165, 166, 167, 174
VIDEO SIGNALS 12, 101, 166
VIDEO SYNC TIPS 181
VIDEOTAPE 8, 99, 102
VIDEO TRACKS 27, 179, 186
VIDEO/CHROMA CIRCUITRY 28, 30
VIDEO/CHROMA PLAYBACK CIRCUITRY 28
VIDEOGRAPHER 1
VIDEOTAPE 47, 48, 49, 99, 107, 110, 111, 112, 138
VIEWFINDER 155
VINYL DISKS 37
VOLTAGE RECTIFICATION 157
VOLTAGE REGULATION 157
VOLTAGE SYNTHESIZER CIRCUITRY 134
VOLTAGE TRANSFORMATION 157

W

WAVEFORM ANALYZER 22
WAVEFORMS 23, 24, 25, 32, 205, 206, 207, 209
WAVESHAPE 104
WHITE PEAK FREQUENCIES 178
WHITE STREAKS 14
WHITE TRAILING HORIZONTAL LINES 46
WIDTH 186
WINDING 63
WIRES 50
WORM GEAR 54, 94
WORM GEAR PULLEY 80
WORN 1, 12
WORN BELTS 107
WORN DRIVE BELTS 103
WORN HEADS 7, 8
WORN IDLER TIRES 56
WORN LOADING BELT 86
WRITING SPEED 198

Y

Y/C CIRCUITS 185

Z

ZENER DIODES 91
ZOOM 82, 155

Exploring the World of SCSI
by Louis Columbus

Focusing on the needs of the hobbyist, PC enthusiast, as well as system administrator, *The World of SCSI* is a comprehensive book for anyone interested in learning the hands-on aspects of SCSI. It includes how to work with the Logical Unit Numbers (LUNs) within SCSI, how termination works, bus mastering, caching, and how the various levels of RAID provide varying levels of performance and reliability. This book provides the functionality that intermediate and advanced system users need for configuring SCSI on their systems, while at the same time providing the experienced professional with the necessary diagrams, descriptions, information sources, and guidance on how to implement SCSI-based solutions. Chapters include: How SCSI Works; Connecting with SCSI Devices; and many more.

Dictionary of Modern Electronics Technology
Andrew Singmin

New technology overpowers the old everyday. One minute you're working with the quickest and most sophisticated electronic equipment, and the next minute you're working with a museum piece. The words that support your equipment change just as fast.

If you're looking for a dictionary that thoroughly defines the ever-changing and advancing world of electronic terminology, look no further than the Modern Dictionary of Electronics Technology. With up-to-date definitions and explanations, this dictionary sheds insightful light on words and terms used at the forefront of today's integrated circuit industry and surrounding electronic sectors.

Whether you're a device engineer, a specialist working in the semiconductor industry, or simply an electronics enthusiast, this dictionary is a necessary guide for your electronic endeavors.

Communication
500 pages • paperback • 7-3/8" x 9-1/4"
ISBN 0-7906-1210-0 • Sams 61210
$34.95

Electronics Technology
220 pages • paperback • 7 3/8 x 9 1/4"
ISBN: 0-7906-1164-4 • Sams 61164
$34.95

To order today or locate your nearest Prompt® Publications distributor at 1-800-428-7267 or www.samswebsite.com

Prices subject to change.

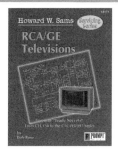

Servicing Zenith Televisions
Bob Rose

Expanding on the HWS Servicing Series, author Bob Rose takes an in-depth look at Zenith TVs, with coverage of manufacturer history, test equipment, literature, software, and parts. A variety of chassis are given a thorough analysis.

Servicing RCA/GE Televisions
Bob Rose

Let's assume you're a competent technician, with a good work ethic, knowledgeable, having the proper tools to do your job efficiently. You're hungry to learn, to find a way to get an edge on the competition—better yet, to increase your productivity, and hence, your compensation.

In *Servicing RCA/GE Televisions*, author Bob Rose has compiled years of personal experience to share his knowledge about the unique CTC chassis. From the early CTC130 through the CTC195/197 series, Bob reveals the most common faults and quickest ways to find them, as well as some not-so-common problems, quirks and oddities he's experienced along the way. From the RCA component numbering system to the infamous "tuner wrap" problem, Bob gives you all you need to make faster diagnoses and efficient repairs, with fewer call-backs—and that's money in the bank!

Troubleshooting and Repair
352 pages • paperback • 8-1/2" x 11"
ISBN 0-7906-1216-X • Sams 61216
$34.95

Troubleshooting & Repair
352 pages • paperback • 8-3/8" x 10-7/8"
ISBN 0-7906-1171-6 • Sams 61171
$34.95

To order today or locate your nearest Prompt® Publications distributor at 1-800-428-7267 or www.samswebsite.com

Prices subject to change.

Exploring Solid-State Amplifiers
Joseph Carr

Modern Electronics Soldering Techniques
Andrew Singmin

Exploring Solid-State Amplifiers is a complete and authoritative guide to the world of amplifiers. If you're a professional technician or a hobbyist interested in learning more about amplifiers, this is the book for you.

Beginning with amplifier electronics: overcoming the effects of noise, this book covers many useful and interesting topics. It includes helpful, detailed schematics and diagrams to guide you through the circuitry and construction of solid-state amplifiers, such as: Transistor Amplifiers; Junction Field-Effect Transistors and MOSFET Transistors; Operational Amplifiers; Audio Small Signal and Power Amplifiers; Solid-State Parametric Amplifiers; and Monolithic Microwave Integrated Circuits (MMICs). Two bonus chapters are devoted to troubleshooting circuits and selecting solid-state replacement parts.

The traditional notion of soldering no longer applies in the quickly changing world of technology. Having the skills to solder electronics devices helps to advance your career.

Modern Electronics Soldering Techniques is designed as a total learning package, providing an extensive electronics foundation that enhances your electronics capabilities. This book covers how to solder wires and components as well as how to read schematics. Also learn how to apply your newly learned knowledge by following step-by-step instructions to take simple circuits and convert them into prototype breadboard designs. Other tospic covered include troubleshooting, basic math principles used in electronics, simple test meters and instruments, surface-mount technology, safety, and much more!

Electronics Technology
240 pages • paperback • 7-3/8" x 9-1/4"
ISBN: 0-7906-1192-9 • Sams: 61192
$29.95

Electronics Basics
304 pages • paperback • 6" x 9"
ISBN: 0-7906-1199-6 • Sams 61199
$24.95

To order today or locate your nearest Prompt® Publications distributor at 1-800-428-7267 or www.samswebsite.com

Prices subject to change.

SMD Electronics Projects
Homer Davidson

Guide to HDTV Systems
Conrad Persson

SMD components have opened up a brand-new area of electronics project construction. These tiny components are now available and listed in many of the electronics mail-order catalogs for the electronics hobbyist. Projects include: earphone radio, shortwave receiver, baby monitor, cable checker, touch alarm and many more. Thirty projects in all!

As HDTV is developed, refined, and becomes more available to the masses, technicians will be required to service them. Until now, precious little information has been available on the subject. This book provides a detailed background on what HDTV is, the technical standards involved, how HDTV signals are generated and transmitted, and a generalized description of the circuitry an HDTV set consists of. Some of the topics include the ATSC digital TV standard, receiver characteristics, NTSC/HDTV compatibility, scanning methods, test equipment, servicing considerations, and more.

Projects
320 pages • paperback • 7-3/8" x 9-1/4"
ISBN 0-7906-1211-9 • Sams 61211
$29.95

Video Technology
256 pages • paperback • 7-3/8" x 9-1/4"
ISBN 0-7906-1166-X • Sams 61166
$34.95

To order today or locate your nearest Prompt® Publications distributor at 1-800-428-7267 or www.samswebsite.com

Prices subject to change.

Home Automation Basics
Practical Applications Using Visual Basic 6
by Tom Leonik, P.E.

This book explores the world of Visual Basic 6 programming with respect to real-world interfacing, animation and control on a beginner to intermediate level.

Home Automation Basics demonstrates how to interface to a home automation system via the serial port on your PC. Using a programmable logic controller (PLC) as a home monitor, this book walks you through the process of developing the home monitor program using Visual Basic programming. After programming is complete the PLC will monitor the following digital inputs: front and rear doorbell pushbuttons, front and rear door open sensors, HVAC systems, water pumps, mail box, as well as temperature controls. The lessons learned in this book will be invaluable for future serial and animations projects!

Electronics Technology
386 pages • paperback • 7-3/8" x 9-1/4"
ISBN 0-7906-1214-3 • Sams 61214
$34.95

Home Automation Basics II: LiteTouch System
James Van Laarhoven

Daunted by the thought of installing a world-class home automation system? You shouldn't be! LiteTouch Systems has won numerous awards for its home automation systems, which combine flexibility with unlimited options. James Van Laarhoven takes LiteTouch to the next level with this text from Prompt® Publications. Van Laarhoven helps to make the installation, troubleshooting, and maintenance of LiteTouch 2000® a less daunting task for the installer, presenting information, examples, and situations in an easy-to-read format. LiteTouch 2000® is a true computer-control system that reduces unsightly switch banks and bulky high-voltage control wiring to a minimum.

Van Laarhoven's efforts make LiteTouch 2000® a product to be utilized by programmers of varying backgrounds and experience levels. *Home Automation Basics II* should be a part of every electrical and security professional's reference library.

Electronics Technology
304 pages • paperback • 7-3/8" x 9-1/4"
ISBN 0-7906-1226-7 • Sams 61226
$34.95

To order today or locate your nearest Prompt® Publications distributor at 1-800-428-7267 or www.samswebsite.com

Prices subject to change.

Complete RF Technician's Handbook, *Second Edition*
Cotter W. Sayre

This is THE handbook for the RF or wireless communications beginner, student, experienced technician or ham radio operator, furnishing valuable information on the fundamental and advanced concepts of RF wireless communications.

Circuits found in the majority of modern RF devices are covered, along with current test equipment and their applications. Troubleshooting down to the component level is heavily stressed. RF voltage and power amplifiers, classes of operation, configurations, biasing, coupling techniques, frequency limitations, noise and distortion are all included, as well as LC, crystal and RC oscillators. RF modulation and detection methods are explained in detail – AM, FM, PM and SSB – and associated circuits, such as AGC, squelch, selective calling, frequency multipliers, speech processing, mixers, power supplies, AFC, multiplexing, cellular and microwave technologies.

Communications Technology
366 pages • paperback • 8-1/2 x 11"
ISBN: 0-7906-1147-3 • Sams: 61147
$34.95

The Right Antenna
Second Edition
Alvis J. Evans

Television, FM, CB, cellular phone, satellite and shortwave signals are available in the air to anyone, but it takes a properly selected and installed antenna to make them useful. With easy-to-understand text and clearly illustrated examples, *The Right Antenna* will give you the confidence to select and install the antenna that meets your needs.

The Right Antenna explains how antennas work and breaks them into TV and FM for discussion. A separate chapter is devoted to interference, and antennas used by hams for antenna band operation. This new second edition also includes chapters on DSS, other satellite antennas, and TV and amplifier antennas.

Also included: How Antennas Work, Selection of Antennas for Specific Needs, Installation of Antennas, Fringe Area and MATV Antennas, How to Identify and Eliminate TV Interference.

Communications Technology
120 pages • paperback • 6 x 9"
ISBN: 0-7906-1152-X • Sams: 61152
$24.95

To order today or locate your nearest Prompt® Publications distributor at 1-800-428-7267 or www.samswebsite.com

Prices subject to change.

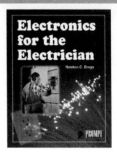

DVD Player Fundamentals
John Ross

Electronics for the Electrican
Newton Braga

DVD Player Fundamentals will cover every aspect of the Digital Versatile Disc starting with features, specifications, hookup, operation, user error, and more. For the professional, specific technical information will include: DVD track structure and disc construction, optical head and lens features and specifications, video signal processing, MPEG-2 technology, audio signal processing, decoding, audio path and reference signals, switch-mode power supply, tracking servo, transverse servo, system control circuits, along with other related topics, including:
- Understanding DVD fundamentals.
- MPEG Technology
- Troubleshooting & Repair

Author Newton Braga takes an innovative approach to helping the electrician advance his or her career. Electronics have become more and more common in the world of the electrician, and this book will help the electrician become more comfortable and proficient at tackling the new tasks required of him or her. Projects and topics include:
- Circuits
- Components
- Fiber Optics
- Troubleshooting Tips

Troubleshooting & Repair
304 pages • paperback • 7-3/8" x 9-1/4"
ISBN 0-7906-1194-5 • Sams 61194
$29.95

Electrical Technology
320 pages • paperback • 7-3/8" x 9-1/4"
ISBN 0-7906-1218-6 • Sams 61218
$34.95

To order today or locate your nearest Prompt® Publications distributor at 1-800-428-7267 or www.samswebsite.com

Prices subject to change.

Troubleshooting & Repair Guide to TV, *Second Ed.*
Engineers of Sams Technical Publishing

The Sams Troubleshooting & Repair Guide to TV is the most complete and up-to-date television repair book available, with tips on how to troubleshoot the newest circuits in today's TVs. Written for novice and professionals alike, this guide contains comprehensive and easy-to-follow basic electronics information, coverage of television basics, and extensive coverage of common TV symptoms and how to correct them. Also included are tips on how to find problems, and a question-and-answer section at the end of each chapter.

This repair guide is illustrated with useful photos, schematics, graphs, and flowcharts. It covers audio, video, technician safety, test equipment, power supplies, picture-in-picture, and much more. *The Troubleshooting & Repair Guide to TV* was written, illustrated, and assembled by the engineers and technicians of Howard W. Sams & Company, who have a combined 200 years of troubleshooting experience.

Troubleshooting & Repair
263 pages • paperback • 8-1/2 x 11"
ISBN: 0-7906-1146-7 • Sams: 61146
$34.95

In-Home VCR Mechanical Repair & Cleaning Guide
Curt Reeder

Like any machine that is used in the home or office, a VCR requires minimal service to keep it functioning well and for a long time. However, a technical or electrical engineering degree is not required to begin regular maintenance on a VCR. *The In-Home VCR Mechanical Repair & Cleaning Guide* shows readers the tricks and secrets of VCR maintenance using just a few small hand tools, such as tweezers and a power screwdriver.

This book is also geared toward entrepreneurs who may consider starting a new VCR service business of their own. The vast information contained in this guide gives a firm foundation on which to create a personal niche in this unique service business.

This book is compiled from the most frequent VCR malfunctions author Curt Reeder has encountered in the six years he has operated his in-home VCR repair and cleaning service.

Troubleshooting & Repair
222 pages • paperback • 8-3/8 x 10-7/8"
ISBN: 0-7906-1076-0 • Sams: 61076
$24.95

To order today or locate your nearest Prompt® Publications distributor at 1-800-428-7267 or www.samswebsite.com

Prices subject to change.

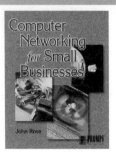

Computer Networking for Small Businesses
John Ross

Guide to Webcams
John Breeden & Jason Byre

Small businesses, home offices, and satellite offices have flourished in recent years. These small and unique networks of two or more PCs can be a challenge for any technician. Small network systems are vastly different from their large-office counterparts. Connecting to multiple and off-site offices provides a unique set of challenges that are addressed in this book. Topics include installation, troubleshooting and repair, and common network applications relevant to the small-office environment.

Webcams are one of the hottest technologies on the market today. Their applications are endless and cross all demographics. Digital video cameras have become an increasingly popular method of communicating with others across the Internet. From video e-mail clips to web sites that broadcast people's lives for all to see, webcams have become a way for people to battle against the perceived threat of depersonalization caused by computers and to monitor areas from anywhere in the world. Projects and topics include picking a camera, avoiding obstacles, setup and installation, and applications.

Available in December

Communication
368 pages • paperback • 7-3/8" x 9-1/4"
ISBN 0-7906-1221-6 • Sams: 61221
$39.95

Video Technology
320 pages • paperback • 7-3/8" x 9-1/4"
ISBN 0-7906-1220-8 • Sams: 61220
$29.95

To order today or locate your nearest Prompt® Publications distributor at 1-800-428-7267 or www.samswebsite.com

Prices subject to change.